GW00702922

# Health and Safety in Youth and Community Work

## A Pocket Guide

**Doug Nicholls**

## HEALTH AND SAFETY IN YOUTH AND COMMUNITY WORK – A POCKET GUIDE

Russell House Publishing Limited

First published in 1997 by:
Russell House Publishing Limited
4 St. George's House
The Business Park
Uplyme Road
Lyme Regis
Dorset DT7 3LS

© Doug Nicholls

All rights reserved. No part of this publication may be
reproduced, stored in a retrieval system or transmitted in any
form, or by any means, electronic, mechanical, photocopying,
recording or otherwise, without the prior permission of the
copyright holder and the publisher.

British Library Cataloguing-in-Publication Data:
A catalogue record for this manual is available from the British
Library.

ISBN: 1-898924-03-1

Text design and layout by: Jeremy Spencer, London

Printed by: Cromwell Press

# Contents

About the author       v

Acknowledgements       vi

**Section One: Introduction**       1

**Section Two: Policy and practice**       4
- *Health and safety support check-list*       5
- *Model safety and security survey*       10
- *Health and safety risk check-list*       14
- *Model accident report form*       24

**Section Three: The eight 'Rs'**       29
- Regulations – the main legislation outlined       29
- Rights       30
- Responsibilities       33
- Risk assessment – a simple structure       35
- Representation       37
  - Before       37
  - After       39
- Representatives       40
- RIDDOR – Reporting of Injuries, Diseases and Dangerous Occurrences Regulations       41
- Rest       43
  - Background       43
  - The Working Time Directive       44

**Section Four: Implementing a health and safety policy**       45
- *Model health and safety policy*       46

**Section Five: Stress at work**       48
- Stress inducers in youth and community work       48
- *Stress inducers check-list*       49
- *Model statement on stress*       51
- Considering a case in law       52
- *Stress claim check-list*       54

**Section Six: Violence and aggression at work**       58
- Available support       59
- Points for consideration       61
- *Model statement on violence at work*       64
- *Model incident report form*       66

**Section Seven: Minibus safety**                    69
   General observations                         69
   *Minibus safety check-list*                   71
   Seatbelts                                     73
   The vehicle                                   74
   Responsible enough to drive?                  75
   Licensing arrangements                        77

**Section Eight: Harassment and bullying at work**   80
   *Model policy on sexual harassment*           81
   The effects of sexual harassment              82
   What should we be doing about it?             82
   Grievance procedure                           83
   Sexual harassment – key elements              86
   Bullying                                      86

**Section Nine: Education outdoors, and on trips and
visits**                                             90
   *Model policy statement on educational visits,
      exchanges and trips*                     90
   Approval                                      91
      Staffing ratios                          92
      Parents/guardians                        94
   Parental consent forms                        94
      *Model parental consent form*            96
   Activities deemed hazardous                   98
      Risk assessment                          99
      Specific activities                     100
   Insurance                                    105

**Section Ten: Resources – training, organisations and
publications**                                      110
   The union                                    110
   The employer                                 111
   Key organisations                            112
   Publications                                 118
      Journals and bulletins                  118
      Other resources                         119

**Useful contacts**                                  121

## *About the author*

Doug Nicholls has been General Secretary of the
Community and Youth Workers' Union since 1987 and a
member of the Executive Committee of the General
Federation of Trade Unions since 1995. He has represented
youth and community workers for twenty years. He
currently sits on the TUC's New Unionism Committee, the
JNC and ETS Committees in England and Wales. He
represents the JNC Staff Side on the NYA's Advisory Council.
In 1993 he chaired the NYA's working party to consider the
core elements of youth and community work training.

Prior to being elected CYWU General Secretary, he was a
full-time community worker in Coventry. This followed a
seven-year period in Oxford as a part-time youth worker
and volunteer adult education tutor while working on full-
time research.

His publications include: *Heliotropes Not Really* (1979), *The
Dance of Death* (1979), *Poems for a Change* (1980), *May Day: a
Short History* (1986), *New Life* (1994), *Employment Practice and
Policies in Youth and Community Work* (1995) and an *Outline
History of Youth Work, Community Work and its Union* (1997),
*The Eve of Europe* (1997), *Youth, Community and Play Workers:
Referendum Now, No to a Single Currency* (1997), *Managing
Aggression and Violence: a Model of Legal Compliance, Safe
Working Practices and Good Personal Safety Habits for Staff*,
with Willie More (1997), *The Novels of William Ash, An
Introduction* (TBP) and numerous published articles on youth
and community work, trade unionism and politics.

The author has a special interest in health and safety issues
in this sector.

## Acknowledgements

Thanks go to Lindsey Harvey, Nigel Knight, John Fijalkowski, CYWU Branch members; Larry McCready, Chair of Salaries and Tenure, and John Stevenson, CYWU's Membership Organiser, should be thanked for their support and critical reading of a draft of both texts. John's close reading and advice were particularly appreciated. Thanks also to Sally Lyons at the National Union of Teachers, Carolyn Oldfield at the National Youth Agency , the Royal Society for the Prevention of Accidents, the Health and Safety Executive and Bob Wiskins at the Birmingham Health and Safety Advice Centre all for accessing useful information and publications for me, and to Willy More of Pepar Publications which produces excellent computer interactive health and safety materials for specialist occupation groups. Peter Noble an expert in health and safety in outdoor education and Dane Oliver were kind enough to point me in some useful directions and to read through what is written on the subject here. All remaining text is of course my responsibility.

This is literally a pocket guide, and will hopefully serve as a mobile instant reference to some of the key elements of health and safety at work and some useful sources of information. It is a companion document to *Health and Safety in Youth and Community Work – A Resource Manual* which provides more back up material, goes into more depth and assists training. This larger work also outlines the key elements in establishing a health and safety campaign to bring in improvements over time and provides full model procedures on various aspects. The Pocket Guide serves a bit more like a 'know your rights, instant back up' document. There is one other document you may consider purchasing immediately if you are considering improving practice significantly, this is *Health and Safety Law – A Guide for Union Reps*, published by the Labour Research Department, June 1997, 78 Blackfriars Road, London, SE1 8HF, 0171 928 3649.

Youth and community workers play an increasing role in making young people and adults aware of their legal and employment rights and how to live and work more healthily, however, we have neglected ourselves as workers in this regard. There is no specific literature on health and safety which takes into account our particular occupational culture and the complex diversity of problems that we face. Youth and community workers work in vulnerable, usually unsecured premises and often with groups of young people, or adults, prone to behaviourial variations. We can work in schools, colleges, on the streets, in large community centres, or small village youth clubs. Youth workers educate at the top of mountains, on four week visits to other countries, in residential centres, on the sports field, on the campsite etc.. We work for small voluntary management committees, or huge local authorities. We work with young people, in canoes, at sporting events, abroad, in residential centres, in outdoor activity centres and in minibuses. We take young people for their first experiences of hazardous pursuits. We play activity games with young people. We work where all the community smoke, altogether in one room all at the same time! Above all, we work in an occupation with minimal management training, a lack of support structure and dwindling resources. By comparison with the formal education sector we are much more vulnerable, yet nothing has emerged from the DfEE, or any major umbrella body to give greater security and protection to our work and staff. Therefore, there is some urgency about how we address health and safety.

Because of the diversity of settings in which we work and the complexity of different health and safety issues we encounter, this pocket guide must be considered as general in the sense that it seeks to identify common problems and the underlying laws and principles which apply to all considerations of health and safety. The manual explains things in more detail.

Everyone is responsible for health and safety, but workers and employers have particular legal responsibilities. In youth and community work we have to be aware of a huge range of health and safety laws and procedures. Community centre workers will have to be experts in the Food Act as well as Fire Regulations and dozens of other bits of legislation. Detached workers will have to understand the complex risks of street work and their employers' liabilities. Workers taking young people on a minibus or to an outdoor pursuits centre will face more rules and regulations and hazards. In terms of their own predicament, youth and community workers will encounter daily pronounced workplace stress hazards that need constant assessment through supervision. Like our teaching colleagues, many of us work in dangerously dilapidated and depressing premises which adds to the danger at work. My concern in both of these publications is to provide ideas to raise awareness, explain some legal rights and responsibilities and suggest some preventative strategies to improve health and safety in our work. As always with a subject like this the trade union is seeking to give information to liberate professionals so that they can get on with their educational work, free from danger and harm; but to get to that point we have to take a number of procedures and laws into account.

## Policy and practice

Your contract of employment is made up of a written statement of particulars which will spell out the fundamental contractual terms such as pay, holiday entitlement and so on. But not everything that forms the contract between a worker and their employer is written down in this way. The union may have negotiated certain agreements with the employer which do not appear in every members of staff's contract. The employer may have rules and regulations and policies which it expects you to follow which are part of the employment agreement with the workforce. In addition to the legal entitlements, such as 'the duty of care' that an employer has to ensure that staff are free from harm and have risks to their health at work properly assessed, there are a number of unwritten contractual obligations that do not appear directly in the statement of particulars. So one of the most important things to do is to be aware of all of those policies and procedures that apply to your work situation and what the employers expect of you in relation to various activities, or in this case health and safety requirements. To play the game you need to know what the rules are.

All of these rules are of course negotiable and if they do not exist you can get copies of good practice and negotiate them into place with your employers. If they exist but are out of date or inappropriate, you can negotiate to improve them.

Before you set out on your long term campaign to improve health and safety in your youth and community service/ project you should review the following policies and practices and ensure that all staff have easy access to locally agreed versions of them and regular training about them. This is a non-exhaustive list.

## Health and safety support check-list

|  | Yes | No |
|---|---|---|
| • Are all the health and safety (H&S) laws and regulations available to you | ☐ | ☐ |
| • Is there a health and safety policy within your employing organisation | ☐ | ☐ |
| • Do you know the name and contact of the CYWU representative | ☐ | ☐ |
| • Do you know the name and contact of the employer's H&S representative | ☐ | ☐ |
| • Is there an accident report book at work | ☐ | ☐ |
| • Are the H&S procedures available at work | ☐ | ☐ |
| • Are all staff, volunteers and young people equally trained, informed and aware | ☐ | ☐ |
| • Has your workplace been inspected for risks | ☐ | ☐ |
| • When ................................................................................................... | | |
| • Is there a thermometer in the office | ☐ | ☐ |
| • Do you know the legal thresholds of temperature at work | ☐ | ☐ |
| • When was the electrical equipment last inspected ......................................... | | |
| • How many cubic metres are available to each member of staff............................ | | |
| • Are there hazardous substances on the premises | ☐ | ☐ |
| • Have you been given training in the COSHH regulations | ☐ | ☐ |
| • Have you been given training in RIDDOR | ☐ | ☐ |
| • Is there a fire alarm | ☐ | ☐ |
| • Are fire exits clearly identified | ☐ | ☐ |
| • Even for the visually or aurally impaired | ☐ | ☐ |
| • Are wheelchair users safe in your building | ☐ | ☐ |
| • How do you evacuate the building ................................................................ | | |
| • Is there a first aid box | ☐ | ☐ |
| • Is it adequate | ☐ | ☐ |

| Health and safety support check-list – page 2 | Yes | No |
|---|---|---|
| • Where is it ............................................................................... | | |
| • Are there enough fire extinguishers and are they of the right type | ☐ | ☐ |
| • When was the last fire officer's inspection ..................................... | | |
| • Do you have somewhere to put your clothes and belongings while at work | ☐ | ☐ |
| • Is there a policy on staffing ratios | ☐ | ☐ |
| • What is the policy on working alone ............................................... | | |
| • Are there regulations relating to drugs use | ☐ | ☐ |
| • Have you received training in HIV and Aids related work | ☐ | ☐ |
| • What regulations apply to the administration of prescribed drugs for young people in your care ..................................................................... | | |
| • What is the procedure for dealing with violence at work ................... | | |
| • What protective measures would your employer take if you were threatened by violence by a user ..................................................................... | | |
| • Are all fixtures and fittings safe | ☐ | ☐ |
| • Are there enough toilets and wash basins | ☐ | ☐ |
| • Is there enough ventilation near the photocopier or other equipment | ☐ | ☐ |
| • Have VDU users been made aware of the new legislation | ☐ | ☐ |
| • Have VDU users been given eye tests at the employer's expense | ☐ | ☐ |
| • How often do users take breaks, what does the law say | | |
| • Does the physical working environment need to be improved | ☐ | ☐ |
| • How is work and lack of support contributing to stress ..................... | | |
| • Are the terms of the food hygiene and storage regulations met | ☐ | ☐ |
| • Are new staff provided with H&S training on recruitment | ☐ | ☐ |

| *Health and safety support check-list – page 3* | Yes | No |
|---|---|---|
| • Are staff trained when new risks become evident | ☐ | ☐ |

• What would be considered a serious or imminent danger at this workplace ...............................
...................................................................................................................................................

• What protective equipment is available to youth and community workers in situations of risk

| | Yes | No |
|---|---|---|
| – Mobile phones | ☐ | ☐ |
| – Alarms | ☐ | ☐ |
| – Helplines | ☐ | ☐ |
| – Solicitor's advice lines | ☐ | ☐ |

• What insurance policies exist in case of injury at work ...........................................................
...................................................................................................................................................

| | Yes | No |
|---|---|---|
| • Are working hours and rotas legal | ☐ | ☐ |
| • Is there guidance on hazardous activities | ☐ | ☐ |
| • Is there guidance on outdoor activities | ☐ | ☐ |
| • Is there guidance on minibus driving | ☐ | ☐ |
| • Have you informed management you refuse to drive minibuses with crew seats or without belts | ☐ | ☐ |
| • Do you take another qualified driver on trips over 150 miles | ☐ | ☐ |
| • Is all equipment used by user groups safe | ☐ | ☐ |
| • Does everyone who uses the equipment know how to use it | ☐ | ☐ |

• Are there any particular dangers for women, black, or disabled staff and users ........................
...................................................................................................................................................

| | Yes | No |
|---|---|---|
| • Are you aware of what insurance schemes cover | ☐ | ☐ |
| • Are indemnity forms for parents and young people sufficient for all activities | ☐ | ☐ |
| • Do you have access to the occupational health support services of the local authority | ☐ | ☐ |
| • Is it clear in what circumstances you are responsible on behalf of the employer for upholding their health and safety commitments | ☐ | ☐ |

   • Arrangements for residentials ...............................................................................
   • Arrangements for international exchanges ...........................................................

| *Health and safety support check-list – page 4* | Yes | No |
| --- | --- | --- |
| • Have all hazardous substances and chemicals on your premises been audited, checked, properly stored etc. | ☐ | ☐ |
| • Have staff been trained in the use of such substances | ☐ | ☐ |
| • Are hazardous activities with youth/adult groups clearly specified | ☐ | ☐ |
| • Are parental consent form regulations clear | ☐ | ☐ |
| • Are the terms of insurance policy for you, all staff, various activities clear | ☐ | ☐ |

## Model safety and security survey

Youth and community workers do not always work in the most trouble free areas of the country! Given the mounting problems faced by schools it is perhaps high time that we took a leaf out of the formal education book and started to assess the security and risks of our premises and working areas.

Using the survey devised by the DfEE for Schools (see *Improving Security in Schools*) I offer the following as a template for assessing risks in the general environment, the client group and the centre where we work. A locally adapted version of this would then be a tool in general risk assessment with particular regard to personal safety and a basis for health and safety committee discussions. Specific local criteria under each heading may need to be adopted as would ongoing data collecting procedures. Readers may find a more sensitive gauge to calibrate the risks than the simple 0-5 tick box I have adopted. After this assessment of the contextual situation and vulnerability of staff, a more detailed check-list could then be considered, also given below.

## Model safety and security survey

|  | 0 | 1 | 2 | 3 | 4 | 5 |  |
|---|---|---|---|---|---|---|---|
| | Low Risk | ☐ | ☐ | ☐ | ☐ | ☐ | ☐ High Risk |

**Incidence of crime in the immediate neighbourhood in the last twelve months**

| | | | | | | |
|---|---|---|---|---|---|---|
| Theft/burglary – No. of incidents | ☐ | ☐ | ☐ | ☐ | ☐ | ☐ |
| Vandalism – Extent and frequency | ☐ | ☐ | ☐ | ☐ | ☐ | ☐ |
| Drugs/solvent abuse | ☐ | ☐ | ☐ | ☐ | ☐ | ☐ |
| Alcohol abuse | ☐ | ☐ | ☐ | ☐ | ☐ | ☐ |
| Public disorder – Cases reported | ☐ | ☐ | ☐ | ☐ | ☐ | ☐ |
| Attacks on public | ☐ | ☐ | ☐ | ☐ | ☐ | ☐ |
| Attacks on workers | ☐ | ☐ | ☐ | ☐ | ☐ | ☐ |
| Weapons related offences | ☐ | ☐ | ☐ | ☐ | ☐ | ☐ |
| Arson attacks | ☐ | ☐ | ☐ | ☐ | ☐ | ☐ |
| Car damage/theft | ☐ | ☐ | ☐ | ☐ | ☐ | ☐ |
| Insurance company assessment of area | ☐ | ☐ | ☐ | ☐ | ☐ | ☐ |

**Considering your building more closely**

*Buildings*

| | | | | | | |
|---|---|---|---|---|---|---|
| Visibility of centre – overlooked, or unobserved grounds | ☐ | ☐ | ☐ | ☐ | ☐ | ☐ |
| Boundaries, fences and gates – Effectiveness | ☐ | ☐ | ☐ | ☐ | ☐ | ☐ |
| Entrances – Limited or multiple | ☐ | ☐ | ☐ | ☐ | ☐ | ☐ |
|     – Clearly defined with signs | ☐ | ☐ | ☐ | ☐ | ☐ | ☐ |
| Exits – Self shutting doors | ☐ | ☐ | ☐ | ☐ | ☐ | ☐ |
| Control of entrance/reception area – Staffed, controlled | ☐ | ☐ | ☐ | ☐ | ☐ | ☐ |
| Car parking – Well lit and overlooked | ☐ | ☐ | ☐ | ☐ | ☐ | ☐ |
| Condition of buildings – (Make separate schedule of dilapidations) | ☐ | ☐ | ☐ | ☐ | ☐ | ☐ |

*Model safety and security survey – page 2*

|                                                                      | 0 | 1 | 2 | 3 | 4 | 5 |
|----------------------------------------------------------------------|---|---|---|---|---|---|
| Low Risk □ □ □ □ □ □ High Risk                                       |   |   |   |   |   |   |
| Appearance of buildings                                              | □ | □ | □ | □ | □ | □ |
| Detached buildings                                                   | □ | □ | □ | □ | □ | □ |

*Security*
Vulnerability

|                                                                      | □ | □ | □ | □ | □ | □ |
|----------------------------------------------------------------------|---|---|---|---|---|---|
| Places for intruders to conceal                                      | □ | □ | □ | □ | □ | □ |
| Secure – Exit doors                                                  | □ | □ | □ | □ | □ | □ |
| – Windows                                                            | □ | □ | □ | □ | □ | □ |
| – Roof lights                                                        | □ | □ | □ | □ | □ | □ |
| State of – Locks                                                    | □ | □ | □ | □ | □ | □ |
| – Alarms                                                             | □ | □ | □ | □ | □ | □ |
| Awareness of staff with access to keys etc.                          | □ | □ | □ | □ | □ | □ |
| Security of valuable equipment                                       | □ | □ | □ | □ | □ | □ |
| State of fire precautions                                            | □ | □ | □ | □ | □ | □ |
| General community attitude to centre or project                      | □ | □ | □ | □ | □ | □ |
| Security measures                                                    | □ | □ | □ | □ | □ | □ |
| Is the centre part of a community watch scheme                       | □ | □ | □ | □ | □ | □ |
| Commitment within neighbourhood to 'looking out' for the centre      | □ | □ | □ | □ | □ | □ |
| Are there any unlocked bins or stores around the building            | □ | □ | □ | □ | □ | □ |
| State of security lighting                                           | □ | □ | □ | □ | □ | □ |
| Surveillance equipment                                               | □ | □ | □ | □ | □ | □ |
| Fire detector system coverage                                        | □ | □ | □ | □ | □ | □ |
| Marking of property                                                  | □ | □ | □ | □ | □ | □ |
| Cash handling procedures                                             | □ | □ | □ | □ | □ | □ |

*Model safety and security survey – page 3*

|  | Low Risk 0 | 1 | 2 | 3 | 4 | 5 High Risk |
|---|---|---|---|---|---|---|

*Users/staff*

Reliability and conscientiousness of users ☐☐☐☐☐☐

Protective equipment/clothing for staff ☐☐☐☐☐☐

General demeanour of client group ☐☐☐☐☐☐

Incidence of threatening behaviour among client group towards staff in last twelve months ☐☐☐☐☐☐

Incidence of violence among client group towards staff in last twelve months ☐☐☐☐☐☐

Incidence of violence between members of client group in last twelve months ☐☐☐☐☐☐

Incidence of vandalism ☐☐☐☐☐☐

Difficult to police areas in centre ☐☐☐☐☐☐

Closing and locking up procedures ☐☐☐☐☐☐

Staff ratios and timetables ☐☐☐☐☐☐

Lone working policies ☐☐☐☐☐☐

Incident report book entries last twelve months ☐☐☐☐☐☐

After the overall risk assessment as outlined above has been taken, you could then consider things in more details. Assuming that most youth and community workers are still building based, I provide below a general outline of some factors that might form the basis of a more detailed health and safety risk check-list.

## Health and safety risk check-list

*Please comment on each item*

Satisfactory

**Office related non-electrical equipment and furniture**

What is the condition of the item ☐

*Action needed*............................................................................

Are there any loose floor coverings ☐

*Action needed*............................................................................

Are there any trailing wires ☐

*Action needed*............................................................................

Are there any jagged or broken edges ☐

*Action needed*............................................................................

Is there room between items ☐

*Action needed*............................................................................

Are floors slippery when wet ☐

*Action needed*............................................................................

Are dangerous tools securely kept ☐

*Action needed*............................................................................

Is stackable furniture safe ☐

*Action needed*............................................................................

Are windows and door glass appropriate ☐

*Action needed*............................................................................

Are stored items secure ☐

*Action needed*............................................................................

*Health and safety risk check-list – page 2*

| | Satisfactory |
|---|---|
| Are shelves secure | ☐ |

*Action needed*..................................................................................................

What is the state of general cleanliness      ☐

*Action needed*..................................................................................................

**Kitchen**
(Should be a non-smoking area)

Are tea towels and towels regularly laundered      ☐

*Action needed*..................................................................................................

General state of cleanliness      ☐

*Action needed*..................................................................................................

Are hot water, soap, nail brush and detergent supplies adequate      ☐

*Action needed*..................................................................................................

Are separate personal cleaning sinks provided      ☐

*Action needed*..................................................................................................

Are facilities for washing and drying crockery and utensils adequate      ☐

*Action needed*..................................................................................................

What is the general state of main food preparation and storage equipment      ☐

*Action needed*..................................................................................................

Is there a cleaning rota      ☐

*Action needed*..................................................................................................

Are there kitchen rules      ☐

*Action needed*..................................................................................................

How is waste dealt with      ☐

*Action needed*..................................................................................................

*Health and safety risk check-list – page 3*

**Satisfactory**

Are cutting machines properly guarded ☐

*Action needed*......................................................................

Record training in food hygiene for staff ☐

*Action needed*......................................................................

Is there a display of freezer dates ☐

*Action needed*......................................................................

Are frozen items date recorded ☐

*Action needed*......................................................................

Are cleaning materials properly stored ☐

*Action needed*......................................................................

What dangers are there to younger children ☐

*Action needed*......................................................................

What dangers/access is there for disabled users ☐

*Action needed*......................................................................

**Toilets**

Are there enough ☐

*Action needed*......................................................................

Are all urinals, WCs, basins etc. uncracked and securely fitted ☐

*Action needed*......................................................................

Provision for disposal of sanitary towels ☐

*Action needed*......................................................................

Are water and washing agent supplies and drying facilities adequate ☐

*Action needed*......................................................................

*Health and safety risk check-list – page 4*

**First aid**

Satisfactory

Is the first aider adequately trained ☐

*Action needed*....................................................................................................................

Is the first aid box obviously located and signed ☐

*Action needed*....................................................................................................................

Is the first aid box adequately stocked ☐

*Action needed*....................................................................................................................

When was it last replenished ☐

*Action needed*....................................................................................................................

Are accident forms available ☐

*Action needed*....................................................................................................................

Are there rules for safe conduct and avoiding accidents ☐

*Action needed*....................................................................................................................

Is a qualified first aider on the premises at all times ☐

*Action needed*....................................................................................................................

**Electrical**

Is there a rota for inspecting equipment ☐

*Action needed*....................................................................................................................

When was the last official inspection undertaken by a qualified electrician ☐

*Action needed*....................................................................................................................

Is all equipment, cables, fuses, sockets, plugs etc., in good working order ☐

*Action needed*....................................................................................................................

*Health and safety risk check-list – page 5*

Satisfactory

Are any sockets overloaded ☐

*Action needed*................................................................................

Are junction boxes and mains supply boxes secure ☐

*Action needed*................................................................................

Is equipment used by users connected via a power breaker ☐

*Action needed*................................................................................

**Fire precautions**

Are fire regulations clearly displayed ☐

*Action needed*................................................................................

Are fire procedures known to all staff and users ☐

*Action needed*................................................................................

Are all fire exit signs properly and clearly displayed ☐

*Action needed*................................................................................

Are all fire exits in working order ☐

*Action needed*................................................................................

Are all exits always unobstructed ☐

*Action needed*................................................................................

Are fire extinguishers in working order ☐

*Action needed*................................................................................

Are all upholstery and curtains etc., fire proof ☐

*Action needed*................................................................................

Is furniture and equipment made of safe materials that will not emit poisonous fumes in case of fire ☐

*Action needed*................................................................................

*Health and safety risk check-list – page 6*

| | Satisfactory |
|---|:---:|
| Are smoke detectors fitted | ☐ |

*Action needed.....................................................................................*

Is fire sprinkler system if installed working ☐

*Action needed.....................................................................................*

Are fire drills held ☐

*Action needed.....................................................................................*

When was last fire officer's inspection ☐

*Action needed.....................................................................................*

Is there emergency lighting ☐

*Action needed.....................................................................................*

Are illuminated signs all working ☐

*Action needed.....................................................................................*

**Ventilation and heating**

Are exposed surfaces safe to touch and clean ☐

*Action needed.....................................................................................*

Can the electrical system be overloaded ☐

*Action needed.....................................................................................*

Is fuel storage secure from interference ☐

*Action needed.....................................................................................*

Are any portable systems and gas bottles properly stored and kept ☐

*Action needed.....................................................................................*

Is all ventilation adequate ☐

*Action needed.....................................................................................*

*Health and safety risk check-list – page 7*

**Satisfactory**

When was the last inspection of the system ☐

*Action needed*......................................................................................................

Are legal temperatures always maintained ☐

*Action needed*......................................................................................................

**Building – inside**

Display of health and safety regulations ☐

*Action needed*......................................................................................................

Smoke/fire other detectors working ☐

*Action needed*......................................................................................................

Security systems operational and effective ☐

*Action needed*......................................................................................................

Door openings safe at all times ☐

*Action needed*......................................................................................................

State of floors – even, not slippery ☐

*Action needed*......................................................................................................

Access for disabled persons ☐

*Action needed*......................................................................................................

Fire officers and public entertainments licence and environmental health
requirements complied with ☐

*Action needed*......................................................................................................

Adequate lighting ☐

*Action needed*......................................................................................................

Adequate cleaning arrangements ☐

*Action needed*......................................................................................................

*Health and safety risk check-list – page 8*

|  | Satisfactory |
|---|:---:|
| Adequate storage | ☐ |

*Action needed*......................................................................................................

| Are all load bearing walls, pillars, platforms etc. safe | ☐ |
|---|:---:|

*Action needed*......................................................................................................

| Sufficient litter collecting receptacles | ☐ |
|---|:---:|

*Action needed*......................................................................................................

| State of staircases, banisters, handrails, lighting | ☐ |
|---|:---:|

*Action needed*......................................................................................................

| State underneath stage/staircases | ☐ |
|---|:---:|

*Action needed*......................................................................................................

| Overhead dangers in building | ☐ |
|---|:---:|

*Action needed*......................................................................................................

**Building – outside**

| Access for emergency vehicles | ☐ |
|---|:---:|

*Action needed*......................................................................................................

| State of steps, ramps, stairs, handrails, entrances | ☐ |
|---|:---:|

*Action needed*......................................................................................................

| State of external fittings, guttering, pipes, aerials, bricks, slates, glass, frames, storm covers, tiles etc. | ☐ |
|---|:---:|

*Action needed*......................................................................................................

| Is all piping clear | ☐ |
|---|:---:|

*Action needed*......................................................................................................

| Does litter gather in any area | ☐ |
|---|:---:|

*Action needed*......................................................................................................

*Health and safety risk check-list – page 9*

| | Satisfactory |
|---|:---:|
| Security of external facilities waste disposal/power points/taps etc. | ☐ |

*Action needed*........................................................................................................

**Around the building**

State of paths, fences, car park, etc.      ☐

*Action needed*........................................................................................................

Safety for pedestrians      ☐

*Action needed*........................................................................................................

Safety for drivers      ☐

*Action needed*........................................................................................................

Overhanging projections      ☐

*Action needed*........................................................................................................

Lighting      ☐

*Action needed*........................................................................................................

Hazardous wastes or rubbish      ☐

*Action needed*........................................................................................................

Requirements of planning authority      ☐

*Action needed*........................................................................................................

Regular maintenance      ☐

*Action needed*........................................................................................................

Obstructions/protrusions      ☐

*Action needed*........................................................................................................

Signposts to entrances      ☐

*Action needed*........................................................................................................

## Model accident report form

All premises should have readily accessible accident report forms and should ensure that copies are taken on any trips and excursions. A model form is given below.

It is suggested that sections 1 – 11 should be completed and returned immediately. The remaining sections may require further investigation and should be submitted as soon as relevant details have been gathered.

# Model accident report form

1. **Occupant**

   Which department controls the building/site where the accident happened? ...............................

   ........................................................................................................................................................

2. **Scene of accident**

   Type of site/building ........................................................................................................................

   Address ..............................................................................................................................................

   Number .............................................................................................................................................

   Precise location within building/site where accident occurred ...............................................

   ........................................................................................................................................................

   Time of accident ................................................................................ am/pm  (delete as appropriate)

3. **Injured person**

   Age .....................................................................................................................................................

   Male/female ......................................................................................................................................

   Able bodied/disabled ......................................................................................................................

   If disabled, please specifiy............................................................................................................

   Occupation ........................................................................................................................................

   Employing department ....................................................................................................................

   Employer's address .........................................................................................................................

   ........................................................................................................................................................

   Length of service in this post .......................................................................................................

   If the injured person was a user of facilities, please indicate (e.g. Club member, etc.) ...............

   ........................................................................................................................................................

   Had parental consent forms for this person been issued ......................................................

   Had parental consent forms for this person been returned ...................................................

*Model accident report form – page 2*

**4. Accident**

Date .............................................................................................................................

Time .............................................................................................................................

Details of the accident, what was the person doing at the time, what actually happened .........

.............................................................................................................................

.............................................................................................................................

.............................................................................................................................

.............................................................................................................................

**5. Machinery/equipment**

Was there any machinery or equipment involved          Yes ☐   No ☐

Was it operating at the time          Yes ☐   No ☐

What part of the machinery or equipment caused the injury ...............................................

.............................................................................................................................

**6. Injury**

Death .............................................................................................................................

Injury was serious .............................................................................................................

Asphyxiation ....................................................................................................................

Electric shock ...................................................................................................................

Poison .............................................................................................................................

Fall .............................................................................................................................

Other .............................................................................................................................

Describe injury .................................................................................................................

.............................................................................................................................

.............................................................................................................................

.............................................................................................................................

*Model accident report form – page 3*

7. **Scene of accident**

   Describe level of supervision available at the time of the accident ...........................................

   At the time of the accident was a first aider on hand      Yes ☐   No ☐

   Who was called to the accident
   | | | |
   |---|---|---|
   | Qualified doctor | Yes ☐ | No ☐ |
   | Ambulance | Yes ☐ | No ☐ |
   | Fire Brigade | Yes ☐ | No ☐ |
   | Police | Yes ☐ | No ☐ |

   Any other person called ...........................................................................................................

8. **After the accident the injured person**

   Left work early      Yes ☐   No ☐

   Was going to see own doctor later      Yes ☐   No ☐

   Went to hospital      Yes ☐   No ☐

   Remained in hospital      Yes ☐   No ☐

   Remained at work      Yes ☐   No ☐

   Name and location of hospital ...................................................................................................

   ..........................................................................................................................................

9. **Work day. At what time did the injured person**

   Start work/activity ...................................................................................................................

   Finish work/activity ..................................................................................................................

   Expect to finish ......................................................................................................................

10. **Any other loss or damage to property as a result of the accident. Specify** ...................................

    ..........................................................................................................................................

*Model accident report form – page 4*

**11. Declaration. To the best of my knowledge the above information is correct.**

Date accident first reported ..........................................................................................

Time accident first reported ..........................................................................................

This form submitted to ..................................................................................................

Signature of person completing form ..........................................................................

Position in organisation ................................................................................................

Name (block capitals) ....................................................................................................

Department ......................................................................................................................

Phone number ................................................................................................................

**12. Describe the cause of the accident**

Were the activities authorised ......................................................................................

Were any special conditions prevailing ......................................................................
............................................................................................................................................
............................................................................................................................................

Were there any hazards in relation to the scene of the accident ............................
............................................................................................................................................
............................................................................................................................................

**13. Preventative action**

What action has been taken to avoid future recurrence ............................................
............................................................................................................................................
............................................................................................................................................
............................................................................................................................................

**14. Further action**

Are any further measures needed regarding the working environment, work system or
individuals concerned ....................................................................................................
............................................................................................................................................
............................................................................................................................................

*Model accident report form – page 5*

### 15. Witness/es

Please give full details of any witness/es

| *Name* | *Address* | *Phone number* |
|--------|-----------|----------------|
| .................. | ........................................................... | .............................. |
| .................. | ........................................................... | .............................. |
| .................. | ........................................................... | .............................. |
| .................. | ........................................................... | .............................. |
| .................. | ........................................................... | .............................. |

### 16. Parents/guardians

If accident involved young person/people

| *Name* | *Address* | *Phone number* |
|--------|-----------|----------------|
| .................. | ........................................................... | .............................. |
| .................. | ........................................................... | .............................. |

### 17. Supervisor/s at the time of the accident ........................................
..................................................................
..................................................................

### 18. Claim

Is any party likely to be claiming as a result of this accident

### 19. Signatures

.......................................................................... Date ......................................

.......................................................................... Date ......................................

## Regulation – main legislation outlined

The Health and Safety at Work etc. Act 1974 (HSWA) places duties on employers, self employed people and employees to safeguard working environments and actively create safer workplaces. A number of regulations flow from this together with directives from the European Union and case-law precedence. Employers have a duty of care towards their staff and in a profession and service like ours we all have a duty of care to those entrusted into our care for activities of all sorts. Among those statutes and regulations which youth and community workers will particularly need to be aware of are:

- The Health and Safety at Work Act 1974
- The Management of Health and Safety at Work Regulations 1992
- The Workplace (Health, Safety and Welfare) Regulations 1992
- The Display Screen Equipment Regulations 1992
- The Manual Handling Operations Regulations 1992
- The Provision and Use of Work Equipment Regulations 1992
- The Control of Substances Hazardous to Health Regulations 1994 (COSHH)
- The Electricity at Work Regulations 1989
- The Personal Protective Equipment at Work Regulations 1992
- The Management of Health and Safety at Work (Amendment) Regulations 1994

- The Health and Safety (First Aid) Regulations 1981
  (Revised Code of Practice 1997)
- Trade Union Reform and Employment Rights Act
  1993

Parts of two earlier acts remain in force; the Factories Act
1961 and the Office, Shops and Railway Premises Act 1963.

A fuller account of these statutes is also given in the
manual which complements this pocket guide. It seems a
long list but the underlying principles and practices the laws
lead to are simple and applicable to a wide variety of
situations. The key is the attitude of mind that makes health
and safety and risk assessment a priority when organising
all work and considering your own employment position
and management of others.

Guidance leaflets on all of the directly 'health and
safety' related regulations listed above can be obtained from
the Health and Safety Executive (HSE), PO Box 1999,
Sudbury, Suffolk, CO10 6FS, Tel: 01787 881165, Fax: 01787
313995. The HSE produces excellent materials, including
many free publications. It is advisable to obtain their
materials. In addition, an extremely good and
comprehensive guide to the law and its implications is found
in *Hazards at Work, TUC Guide to Health and Safety*.

## Rights

Employees have a right not to be expected to work in an
environment physically, or mentally injurious or potentially
hazardous to their health. Under the terms of the Trade
Union Reform and Employment Rights Act 1993 employees,
including those expected to work at home can stop work in
the event of serious, or imminent danger arising from the
work that they are doing. Equally of course, a worker has a

duty under legislation to report a fault, or problem which may represent a threat to their health or the safety of others; rights and responsibilities really are usefully integrated in the law governing health and safety.

Employees have a right to refuse to work in a dangerous environment until the employer has made it safe. The employer is breaking a contract of employment by putting employees at risk, so in circumstances where workers are vulnerable they may often have no choice but to withdraw themselves from the danger. Employers must always demonstrate that they have trained their staff in healthy and safe working practice and informed them of health and safety matters, and that they have taken steps to reduce risks to them. The emphasis of the law is on active risk reduction on a regular basis. Managers in all situations, as agents of the employers, must be able to demonstrate with hard, usually documented evidence, that staff engaged in a particular activity, or with a particular responsibility, have been so informed and trained in the health and safety factors involved that they understand all of the hazards and safeguards and accident reporting procedures.

Employers must also be able to demonstrate that they have acted to stop a recurrence of any identified risk and if an employee has been injured as a result of something at work that when they return to work the cause of the injury, say a stressful working environment, has been reduced. It becomes an employee's right to expect that the employer will enact their legal duty of care towards them in this way. Often when workers return from a period of illness, especially if it is stress related, they are able to negotiate a phased return to work, building up their hours each week over a period until they are able to resume full duties without further risk to their health. The employer must actively demonstrate efforts to avoid the repeat of the stress factors.

A worker has the right within any employing organisation and as a member of a collective of workers, through a trade union, to be fully consulted on all aspects of health and safety in the workplace and to be provided with the necessary information required to make this consultation meaningful. New working practices must be subject to a thorough risk assessment. Question, for example, the following if the line manager asks a member of staff to undertake additional duties with another youth centre down the road:

- Will it involve lone working
- Will it involve travel in unsafe territory
- Will it involve managing a building with new risks
- Will the extra work affect the worker's stress levels

Rights of course have to be exercised, they do not apply automatically. This is where the CYWU comes in. Staff have a right to ensure that their union officers are trained in health and safety matters, that they are able to inspect all workplaces where their members work, to negotiate improvements to them and represent individuals when they need representation.

Workers have a right to claim compensation against their employers if they are injured or made ill as a result of demonstrable negligence by that employer. Employees therefore have a right to pursue such a claim to the courts with the assistance of a union. CYWU can provide a free half hour consultation with a solicitor prior to progressing a claim, if required.

If a worker is incapacitated as a result of an injury at work, and off work for a long time, the employer may request that they undertake a medical examination with a doctor of their choosing and that worker then has a right of access to their medical records. Workers have a right under the Medical Records Act (1994) to see and challenge any report prepared about them.

If a worker is hospitalised for a period of 5 days or more they are entitled to hospitalisation benefit from CYWU, if of course they are a member!

All employees have a right to have their working environment and their response to the pressures and risks at work assessed. Additionally there is a right to have a union representative inspect any workplace or work area.

## Responsibilities

Everyone has a shared concern and responsibility for health and safety at work. In the terms of the Health and Safety at Work Act 1974 it is quite clear that "it shall be the duty of every employer to ensure, so far as is reasonably practicable, the health, safety and welfare at work of all his (sic) employees." The employer is required by law to create and sustain safe working environments and practices by whatever means are appropriate – training, safety equipment, supervision, disciplinary procedures and incentives. In a small voluntary organisation, or very often within the local authority, a youth and community worker will be, as a full-time worker particularly, the key representative of the employing organisation. In these circumstances they should therefore ensure that they have produced a health and safety policy for the organisation. The manual provides a full model procedure and campaigning strategy. The Health and Safety Commission's publication *Advice to Employers* is an essential document in this regard. This should definitely be an item on every youth and community centre bookshelf.

Just as youth and community workers should expect their employers to take steps to improve their working environment and protect them from unsafe occurrences and hazards, so the youth and community worker must fully co-operate with the employer in bringing this about. As a front-line service provider they will be expected to have a high level of consciousness about health and safety matters and a contractual responsibility to ensure that the employers' policies and procedures are adhered to in the interests of public safety and those of other employees.

The legislation recognises that employees have obligations too; "to take reasonable care for their own and others' safety and to co-operate with employers so far as is necessary to enable them to carry out their own obligations."

For youth and community workers this is all very important. In a large community centre all of the full and part-time and many of the voluntary staff may be in membership of the union and all will be working directly for, let's say the local authority. If the centre manager is required by the employer to issue an instruction that part-time workers check the brackets on fire extinguishers before each session, this must be carried out. If a part-time worker does not do this and an extinguisher with a faulty bracket falls and breaks the hand of a young person, the local authority will be liable to a claim of negligence from the young person's parents. The part-time worker in this case would be the negligent agent of the local authority and having failed to carry out a very reasonable instruction would be disciplined if not dismissed while the local authority's insurance company haggled over the terms of the compensation to be paid to the young person. The local authority may also want to claim that the centre manager failed to check that the order given had been carried out. If the extinguisher fell on the centre manager's hand instead of the young person's and broke it, he/she may still pursue a claim against the authority if they could prove that, as a result of some failing further up the line the bracket had not been fixed, or the instruction had not been carried out.

We must all be health and safety conscious, we must all seek to assess risks. If we perceive risks it is part of our legal and of course moral responsibility to report them clearly and record such reports properly in writing. We must work to remove the risks and leave the employer in no doubt as to what the risk is and what we expect to be done about it. If the employer fails to respond, they fail in their duty of care and can be sued.

Building based workers in charge of health and safety will have a variety of specific duties set out in the Act, in particular Section 4.

The most important right is to have been trained and regularly updated on health and safety matters at the employers expense, and to feel confident that the union representative for health and safety has been similarly properly coached.

## Risk assessment - a simple structure

This is the mechanism for managing and improving health and safety at work. The management of Health and Safety at Work Regulations 1992 make it a requirement for employers to undertake risk assessments of potential hazards at work that could harm employees. Workers will very often be delegated the responsibility for undertaking some risk assessments for their employer; in larger establishments it should be the responsibility of specialist officers within the local authority.

Risk assessment can apply to any situation from the state of your building to planning a youth expedition to the Himalayas. If the mechanism is not being regularly used, the employer is very likely to be acting illegally. This also applies to any equipment that the employer has authorised workers to use at home – printers or computers for example.

The structure of risk assessment is simple and effective. It can be applied as an approach to assessing the office equipment in a centre, or to assessing a workers' stress levels. It is applicable to a wide range of environments from the safety of an indoor climbing wall to the potential dangers of the detached team's visit to the scene of last weeks riot. Employers need to be able to demonstrate that they have undertaken proper risk assessment in relation to all those factors that could be a hazard to health. Their risk assessment reports should be written, filed and available in the workplaces or to the individuals to which they apply. The employer and staff responsible for carrying out the health and safety responsibilities must be able to demonstrate that they have, for example, ensured that there are no oily patches on the climbing wall, and that the detached team have suitable clothing and communication equipment. The structure of risk assessment is as follows:

1 Identify hazards
2 Decide who might be harmed and how
3 Assess the risks and take appropriate action to remove them or reduce them as far as possible
4 Record the findings
5 Check the risks regularly to assess further preventative measures

At each stage, workers will need to ensure that all relevant factors are taken into account – the qualifications and experience of the person doing the assessment are important for example. The accuracy of the findings at each stage can be contested by the individual concerned, or their representative can submit an alternative report if they do not agree with them. The frequency of follow-up checks may be critical. Always ask – is the employer doing everything it can to reduce the risk to health? Are the precautions agreed reasonable and effective? Is there evidence to show that a proper check was made? Two very helpful publications on this matter are: *Five Steps to Risk Assessment* and *Five Steps to Successful Health and Safety Management*, both published free by the Health and Safety Executive.

An unacceptable risk to a youth and community worker can be the potential threat of violence from an armed and disturbed local resident, the faulty plug, the accumulation of stress at work, the bullying manager, the dilapidated building, or the three hundred young people to one worker at the disco.

As well as risk assessment undertaken by the employers representative, ensure that health and safety checks are made on a co-operative basis between unit staff, management and the health and safety union rep.

## Representation

### Before

The workplace, the levels of stress and the working environment should all be regularly risk assessed as part of an overall health and safety strategy. Prevention is much better than cure. Staff must always make the employer aware of a potential hazard at work, or of a failing which means that they cannot fulfil their part of the contract to take reasonable steps to ensure a healthy working situation. Always make sure such warnings are recorded in writing and filed. Health and safety inspection is a combination of expertise and judgement, so it is always important to ensure that unions and management assess and document their findings together.

As a result of regular inspection and assessment employers and employees will identify improvements that should be made. In identifying the improvements needed workers should also seek to agree a time-scale for the necessary change to take place. If the employer does not agree that the change is needed, or fails to implement an agreed improvement, staff may require representation for their case. Call in the union. The health and safety officer can make an independent check on the matter. If needed, request that the Health and Safety Executive (HSE) or environmental health be brought in. This step should only be taken after all attempts to resolve the matter between union and management have been tried.

A worker's complaint can be formally logged by their union with their employer. In a local authority setting the representative will usually be able to refer the matter to a health and safety committee which may be able to bring its weight behind the case and chase up the local management to remind them of corporate council policies. Additionally health and safety may be a regular tabled item on the agenda of the local Joint Negotiating Committee (lJNC) between management and unions.

In some situations staff will need to take out an immediate grievance. In more serious situations they may need their representative to argue for an immediate removal from the work until the hazard has been minimised beyond all reasonable risk.

If management are proposing something in the future which a worker thinks may be dangerous, they should take out a grievance because until the grievance is resolved the threatened new development cannot take place and the status quo must remain.

If for example management proposes to install two new filing cabinets in a worker's office next Monday and they know that this would reduce their workspace below the legal minimum of 11 cubic metres for each person permanently occupying a workplace, then they should take out an immediate grievance against this move, because the cabinets cannot be installed until the grievance is resolved. Similarly, if the radiators are due for cutting off and draining in the coldest February of the year, staff can formally complain to stop this happening because it is likely that this would mean the temperature of the workplace would fall below the legal minimum of 16°C where people work sitting down. Note there is no maximum temperature.

### After

Despite a worker's best endeavours and those of management, let us say there is an accident. A worker is injured, or someone in the temporary care of a youth worker or in their employment is injured, or the worker is made ill. Here is the advice for instant action:

1. Take a photo of the scene of the accident.
2. Ensure the accident report book is filled in.
3. Report the incident immediately to the line manager.
4. Write the version of events; why the worker was injured, made ill. Give dates and causes.
5. List any witnesses.
6. Consider whether the worker contributed by negligence, or omission to the accident.
7. If a worker wants to see whether they have a case to pursue against an employer for compensation, they

should contact their union caseworker. A worker would ultimately have to be prepared to go to court on this matter. The union would provide a solicitor or barrister to pursue the case. The type of injury will have a price attached to it and the lawyers would argue about the level of contribution that the worker made to the accident.

8  If a worker is hospitalised for 5 days or longer they can claim some expenses (around £20 per day) from CYWU. Contact the national office for a form.

9  When the worker returns to work they should ensure that their employer has removed the cause of their injury/illness and is supporting them to readjust to circumstances and workload. A phased return to work may be necessary and this can be negotiated between the worker, their union rep and their employers.

10 If someone in a worker's care has been injured then clearly they will need to write down a report of their level of responsibility as a manager and not an employer for the hazard that caused the accident, or incidence. They will need to account for the level of evidence they had that staff were aware of dangers, that risk assessments had been undertaken etc.. Similarly, and more importantly, their line manager will have to account for the evidence they had to show that the worker had been trained, made aware and understood the risks involved.

## Representatives

Health and safety representatives are the best category of trade union official to be in terms of being served and protected by law. Their rights to paid time off for training and inspections and negotiations are considerably better than those provided to other sorts of lay trade union officials. Also, if an employer is ever contemplating disciplinary action against a health and safety official, they should approach the full-time official of the union first before it goes ahead. This provides an extra safety valve and an opportunity to avoid unnecessary discipline which could be construed by the union as victimisation.

Each workplace and CYWU branch should have a health and safety representative. They should be fully trained (paid time off is required by law for this), fully conversant with local practice, regularly inspecting workplaces and ensuring that health and safety is on the negotiating and professional agendas. The law requires that they are equipped and trained to do their job as a health and safety rep.

## RIDDOR – Reporting of Injuries, Diseases and Dangerous Occurrences Regulations

New regulations came into force on 1st April 1996. These place a duty on the employer to report and keep a record of certain work related accidents, injuries, diseases and dangerous occurrences. Employers are required to find out about accidents, injuries, diseases or dangerous occurrences which arise from work. Reporting in most cases, relevant to local authority, and regardless of the actual employing organisation, will be to the local authority environmental health department. A record of the report must be kept at the workplace.

If there is a fatality or a major injury, including one resulting from violence at work, this must be reported immediately without delay by telephone. The member of staff must then follow this up with a completed accident report form (F2508). Major injuries according to the HSE's definitions include: fracture other than to fingers, thumbs or toes; amputation; dislocation of the shoulder, hip, knee or spine; temporary or permanent loss of sight; chemical or hot metal burn to the eye or any penetrating injury to the eye; injury resulting from an electric shock or electrical burn leading to unconsciousness or requiring resuscitation or admittance to hospital for more than 24 hours; any other injury leading to hypothermia, heat induced illness or unconsciousness, or requiring resuscitation, or requiring admittance to hospital for more than 24 hours; unconsciousness caused by asphyxia or exposure to harmful substances or biological agent; acute illness requiring medical treatment, or loss of consciousness arising from absorption of any substance by inhalation, ingestion or through the skin; acute illness requiring medical treatment where there is reason to believe that this resulted from exposure to a biological agent or its toxins or infected material.

If the injury results in the worker being away from work for over three days you must complete the form F2508 within ten days, including non-working days.

If something happens which did not lead directly to an injury but could have done and constituted a dangerous occurrence this must also be reported immediately by phone. 21 reportable dangerous occurrences are listed by the HSE in their booklet *Everyone's Guide to RIDDOR 95*.

If an employer is notified by an employee's doctor that they are suffering from a reportable work related disease then the employer must complete another form (F2508A) to the enforcing authority. A list of the main work related diseases in this category is also given by the HSE.

## Rest

### Background

The hours of work and their management are especially important when considering the health of workers. Youth and community workers working full-time should work no more than 32.5 hours a week in 1997, 35 hours a week in 1998 and thereafter 37 hours per week. These hours can be divided across ten sessions worked flexibly in the professional interests of workers. No more than eight evening sessions per fortnight can be worked. These are the main provisions of the *JNC Report for Youth and Community Workers*.

However, given the stresses and strains involved in the unpredictable and intense nature of social education delivery, the conflicting demands of employers, users, and real needs, the overall hours per week identified in JNC is just the start of the story. How those hours are adhered to and how they are distributed are key factors. Assisting the proper management of working time is the new directive on this subject. This makes important provisions which all youth and community workers should ensure they adhere to.

For many years now youth and community workers have taken on more duties and responsibilities and bigger jobs as local services have been pulled apart. This means that it is a crucial moment now to reflect back on our length of working time and work patterns. Eight hours' labour, eight hours' rest, used to be a pivotal Labour Movement demand. Now in a period of mass unemployment, workers seem to want to work all hours available regardless of the consequences to themselves, their families and their communities. Youth and community workers have through massive overwork shown that cuts and unemployment can be hidden. It is hardly surprising that the union's stress and burnout caseload has increased dramatically.

Our campaign to increase job opportunities for qualified students and to ensure the provision of sufficient numbers of trained workers is weakened by members, full and part-time, working so many unpaid, voluntary overtime hours.

With unemployment still at around 5 million in Britain and 18 million in the European Community, overtime is a real blight.

### The Working Time Directive

Many youth and community workers are contravening the new Working Time Directive. This gives a worker the right to:

1  a maximum 48 working week
2  a daily rest period of 11 consecutive hours
3  a minimum weekly rest period of 24 consecutive hours, in addition to the daily rest period, which in principle should include Sunday
4  a rest break where the working day is longer than six hours, and
5  a maximum of eight hours night work in any 24 hour period

This directive is of particular interest in youth and community work when considering the staffing ratios necessary for international exchanges and residential work. How will you ensure that these requirements are maintained and that staff are rested and therefore alert enough at all times to act *in loco parentis* and ensure that those under their supervision are properly cared for? Are youth and community workers always sufficiently rested to undertake their ordinary duties with enthusiasm and skill?

## Section Four
### Implementing a health and safety policy

Every employer employing more than five people must by law write down its policy for their safety and health. This must be shown to a health and safety inspector if requested. For many youth and community organisations, particularly in the voluntary sector, this legal requirement is simply not being fulfilled. Such policies will be organisation specific and excellent advice is available from the Health and Safety Executive. The following publications will help you considerably in constructing your health and safety policy: *Essentials of Health and Safety at Work* HSE 1990 ISBN 0 7176 0716 X, *Safety Policies in the Education Sector* HSE 1994 ISBN 0 7176 0723 2, *Writing a Safety Policy Statement: Advice to Employers* HSC 6 rev. (leaflet). You are strongly advised to read and discuss this material. What is offered below must be regarded as a minimum statement and a general framework.

## Model health and safety policy

**Downtown community centre health and safety policy**

"It shall be the duty of every employer to ensure, so far as is reasonably practicable, the health, safety and welfare at work of all of his employees.

No person shall intentionally or recklessly interfere with or misuse anything provided in the interests of health, safety, or welfare in pursuance of any of the relevant statutory provisions."

*Source: Health and Safety at Work Act, 1974.*

Downtown community centre is committed to creating a healthy and safe working environment for all of its staff and users within the building and surrounding areas and while undertaking activities sponsored and organised by the community centre. Downtown community centre is therefore concerned to ensure that it operates all its activities in full accordance with the relevant health and safety legislation, regulations and official guidance and with the decisions reached between management and CYWU through the local health and safety committee.

Downtown community centre is therefore committed to providing training, supervision, information and equipment to achieve this objective and to maintain a health and safety committee, to undertake regular risk assessments and health and safety inspections and to ensure that health and safety issues are always on staff supervision agendas, management committee agendas and regularly reported to user groups. A notice board and information sheet will be displayed for all those who use the centre. This policy will be regularly reviewed. Downtown community centre also undertakes to review the terms of its insurance, indemnity and liability cover on an annual basis and will seek to fully support staff and volunteers in litigation resulting from an injury sustained from the action of members of the public or users.

Signed ................................................................ Chairperson of management committee.

Signed ................................................................ CYWU health and safety representative.

Date ................................................................

**Responsibilities**

1. Overall responsibility for health and safety within DCC is that of:

   Mr/Mrs/Ms ........................................................................................................................................

   Address ............................................................................................................................................

2. Mr/Mrs/Ms/ ................................................................ is responsible for carrying out this policy within the identified premises coming under the auspices of DCC.

*Model health and safety policy – page 2*

3.     Mr/Mrs/Ms .................................................................. will deputise for this responsibility in her/his absence.

4.     The following managers are responsible for safety in the following designated areas:

    Mr/Mrs/Ms .................................................................. Canteen

    Mr/Mrs/Ms .................................................................. Offices

    Mr/Mrs/Ms .................................................................. Hall etc. etc.

    Those designated above will be responsible for risk assessment, identifying training courses, carrying our safety inspections, monitoring of implementation of policy. They will also report to the health and safety committee of all accident report book entries, all reports coming under the terms set out in RIDDOR and all problems identified in supervision and staff meetings.

5.     All paid staff, volunteers and users will co-operate with designated managers and health and safety representatives in carrying out this policy and specific points. In addition, they will immediately report to the appropriate manager any potential hazard

6.     Consultation between management and CYWU on this policy and general health and safety practice will be conducted by the health and safety committee which will meet at least times per year.

7.     Details of all training opportunities, HSE inspectors, and other specialist advisors and health service agencies together with the decisions of the health and safety committee shall be issued to all members of staff.

8.     Designated first aiders will be ..................................................................

9.     The health and safety committee will draw up house rules for all visiting contractors or other agencies using the premises and it shall also be responsible for ensuring that all lead volunteers and user groups are made aware of health and safety rules.

10.   A health and safety budget will be created.

11.   Rules will be drawn up by the health and safety committee and reviewed annually for: cleanliness, waste disposal, storage and safe stacking, marking of gangways and exits etc, equipment checks and frequency, fire regulations and accident reporting, electrical checks and frequency, special access, use of outdoor areas, security of staff, use of equipment and machinery, hazardous substances and chemicals, all activities and pursuits undertaken by DCC.

## Section Five
### Stress at work

## Stress inducers in youth and community work

In a recent TUC survey of members' health and safety concerns, stress again topped the list. The emphasis of this section is on prevention and a model statement on stress to be incorporated in regular supervision and local negotiations. An excellent training and awareness raising manual has been produced by Russell House Publishing: *Meeting the Stress Challenge: A Training and Staff Development Manual for Social Welfare Managers, Trainers, and Practitioners*, by Neil Thompson, et al., 1996. This is an excellent basis of training in this area for local adaptation by youth and community departments.

Stress factors are now recognised as health and safety hazards as much as dangerous chemicals or other physical dangers. The Health and Safety Executive has a useful summary of some of the main work related stress inducers given in the chart below. These inducers should be seen in the same light as dangerous substances at work, as material things that can damage the body. It is vital that the employer accepts them as such.

## Stress inducers check-list

| | |
|---|---|
| Work characteristics | Conditions defining hazard. (Demands, control and support) |
| Context, organisational function and culture | Poor task environment and lack of definition of objectives. Poor problem solving environment. Poor development environment. Poor communication. Non supportive culture. |
| Role in organisation | Role ambiguity. Role conflict. High responsibility for people. |
| Career development | Career uncertainty. Career stagnation. Poor status or status incongruity. Poor pay. Job insecurity and redundancy. Low social value at work. |
| Decision latitude/control | Low participation in decision making. Lack of control over work. Little decision making in work. |
| Interpersonal relationships at work | Social or physical isolation. Poor relationships with superiors. Interpersonal conflict and violence. Lack of social support. |
| Home/work interface | Conflicting demands of work and home. Low social or practical support at home. Dual career problems. |
| Content/task design | Ill-defined work. High uncertainty in work. Lack of variety or short work cycles. Fragmented or meaningless work. Under utilisation of skill. Continual exposure to client/customer groups. |
| Workload pace – quantitative and qualitative | Lack of control over pacing. Work overload or underload. High levels of pacing or time pressure. |
| Work schedule | Shift working. Inflexible work schedule. Unpredictable work hours. Long or unsocial work hours. |

This useful set of definitions also provides an ideal basis for a stress survey. Give a score of one to five for each factor in all supervision sessions as part of the risk assessment.

These definitions are taken from a research report on stress published by the Health and Safety Executive, (HSE Contract Research Report No. 61/1993, *Stress Research and Stress Management: Putting Theory to Work*, Tom Cox, Centre for Organisational Health and Development, Department of Psychology, University of Nottingham, NG7 2RD). These columns, particularly the one on the right can usefully be considered in relation to the occupations of youth and community work and others within the informal education sector, and considered in relation to remedying what is one of the most destructive aspects of our work, the high stress levels. I believe that given these very commonly identified stress factors, youth and community work has an especially high concentration as an occupation. Without denigrating the stress levels that have clearly escalated in other professional groups, there are some additional features present within youth and community work that contribute significantly to stress; these include questions of status, legal underpinnings of the work, supportive work structures, lines of accountability and task and work objectives.

In this regard the establishment of a baseline policy from which stress reduction strategies can flow is suggested as follows:

## Model statement on stress in youth and community work and stress reduction

1.  This organisation recognises occupational stress as a serious potential health and safety problem that must be systematically minimised and monitored. It is recognised that stress is not an individual illness that can be cured by counselling, but that it is a complex phenomena that can cause illness in individuals both psychological and physical. It is recognised also that various categories of staff may be subject to diverse forms of different stress risk: detached workers, community centre workers, youth workers, play workers, street workers, lesbian workers, gay workers, disabled workers, women workers and men workers may all face different stress factors depending on context. It will be the intention over time through monitoring to analyse common patterns and devise reduction strategies.

2.  Stress is recognised as resulting from a variety of factors in the working environment that may include the following: lack of status and reward, ambiguity and conflict in role, unsocial working hours, long hours, conflicting work demands, uncertainties in career development and expectation, high levels of responsibility for people, poor communication, the lack of mechanisms for solving problems, lack of appreciation, isolation at work, low levels of participation in decision making, poor working relationships, continual exposure to individuals and community groups with various and sometimes conflicting demands. It is further recognised that stress manifests itself as a result of some or all of these factors at various levels of intensity in different individuals at different stages of their career or life.

3.  It is therefore a requirement of regular monthly supervision that in the assessment of general matters relating to health and safety that consideration will be given to factors likely to cause stress. The potential hazards involved shall be assessed and strategies for remedy will be planned and reviewed at subsequent meetings. Persistent contributory factors will be reported to the health and safety officer and this organisation does not expect workers to be subject to factors which are consistently aggravating stress. Every effort will be made to ensure that users are informed of the appropriate standards of behaviour procedures and equal opportunities policies and workers will not be expected to work within potentially violent, unnecessarily conflictual, or unresolved situations.

4.  It is recognised that clear management expectations, realistic work plans, an ability to seek supervisory support, in-service training and the regular identification of problems are helpful devices in stress reduction and staff and the employing organisation have a responsibility to make these effective. Clear purpose in all of the varied duties and responsibilities of youth and community workers is essential and this will be reviewed regularly through supervision. The identification of critical success and failure factors will be integral to stress monitoring through supervision.

5.  It is the intention of this general statement to create an environment in which all staff feel able to:
    *   Identify contributory factors towards their experience of stress.
    *   Agree strategies of intervention to reduce it.
    *   Become informed of ways of reducing and identifying problems in the future.
    *   Monitor and evaluate these strategies.
    *   Obtain management support for stress reduction within an overall concern for the health and safety and well being of staff.

## Considering a case in law

One of the best incentives to prevention, regrettably, is taking a case for injury to health against the employer. If you think that there may be a member of your branch who deserves compensation for the harm caused to them by the employer's failure to care for them and relieve the stress inducers at work, please consider the list of legal points in such cases. If the local situation feels as if it is fitting the points made, please contact CYWU directly at national office and we will be pleased to give further consideration to the matter. Taking a legal case requires rigorous examination of the facts and evidence and the union would have to prove at least four main points. These are:

- that the member was actually mentally/physically injured as a result of the work stress
- that the employer could have foreseen the stress levels and therefore prevented them
- that the employer failed to take the necessary reasonable steps to prevent the stress factors at work
- that there were actually substantial work based stress factors.

In assessing whether a member may have a good case to pursue to the courts for negligence and damages against the employer, close consideration will need to be given to the following points. The branch may not be in a position to examine them all closely, but they may assist some initial filtering before referring the case to national. Do not raise any hopes, seek facts. These points may also assist you towards a better understanding of stress and the employers' responsibility.

Clearly, such questioning can itself be highly stressful and an individual's esteem and memory when suffering or recovering from stress are not at their best. The individual may also still feel to blame. By deploying the best listening and supportive youth and community work skills, the following areas, put here as harsh questions in check-list form, need to be addressed.

## Stress claim check-list

1. Has the member, their parents or grandparents any history of mental illness or sought any previous psychiatric advice from a doctor or psychiatrist? A family history of general mental illness may weaken chances of success or demonstrate a susceptibility to stress which, if the employers were aware of it, could be useful.

2. During the time when the member was suffering work-related stress were there any other stress factors in their lives? Moving home, bereavement, ending a relationship, caring for sick relatives or friends, such things are known stress factors.

3. When did the member first start working in the stressful environment and when did they first complain of it?

4. Did the member clearly and demonstrably request someone in authority to relieve the situation (heavy workload, understaffing, unclear objectives etc.) that was causing their stress?

5. What was it about the work that was causing the stress? Overwork or heavy workload are not in themselves recognised as the only causes of stress. See the attached chart from the Health and Safety Executive.

6. When did the member first experience the stress?

7. What were the symptoms and how did they develop?

8. When did the member first seek medical advice and what was the nature of that advice and diagnosis?

9. Precisely how was their illness or absence from work described to their employer?

10. What steps were taken by the employer to establish whether the member was fit to return to work?

*Stress claim check-list – page 2*

11. What action did the employer take on the return to work to alleviate the factors that had caused illness in the first place?

12. What general steps did the employer take to make work less stressful?

13. If the illness recurred why were the steps the employers took after the first occasion insufficient?

14. After being told that the member was ill because of work-related stress, what steps did the employer take to prevent recurrence and were they alerted to the development of symptoms that suggested recurrence was likely?

15. Did any other workers in a similar position suffer similarly, or were others aware of the member's situation and what did they do or say about it?

16. Could the employer have realistically done something about it?

17. What was the extent of absence of work and how else did the stress affect the member?

18. Can we clearly describe and prove most aspects of the working environment that we are considering with the member? Could you demonstrate, for example, growing demands on the service, lack of management support, difficulty of work, demands of stressful work judgements, member's frustration at lack of resourcing and inadequate management responses. Remember courts require evidence and witnesses not just opinions.

19. The nature of the member's resilience and strength of character is an issue and some way of describing this in general terms would need to be considered.

The nature of the member's resilience and strength of character is an issue, so a way of describing this in general terms would need to be considered.

Having talked a situation through with a member who has clearly been harmed by their employer's negligence and considered that there may be something worth pursuing, discuss with the member the possibility of seeking further union advice with a view to taking the matter further in order to get them compensation. If the member had tripped over a wire at work and broken a leg, they would have no hesitation in seeking compensation from the employer if the wire was in the wrong place at the time and through no fault of theirs. The situation is the same with stress and questions of competence and pride should not come into it.

Assure the member that they are in control of the process of taking the matter further. Stress often results from a lack of a sense of control over your personal situation, so it is vital that the union is seen as a controllable friend working at their pace. However, if the case is ultimately pursued either through a local grievance or a court case the member will need to be prepared to 'sweat it out' and drag over old and painful coals.

## Further reading

Unfortunately some of the best resources are phenomenally expensive, or available only in North America, or to the members of various trade unions. There is nothing that has been written in Britain of any use in remedying the problem specifically for youth and community workers. One publication aimed at social work is extremely useful for youth and community departments and provides a very valuable training tool *Meeting the Stress Challenge: A Training and Staff Development Manual for Social Welfare Managers, Trainers, and Practitioners*, by Neil Thompson, et al., 1996. One article from overseas well worth getting through the local article research scheme is *Burn Out in the Public Interest Community* by William L. Bryan, 1980, The Northern Rockies Action Group Inc. This looks at the effects of stress on community workers in Canadian community projects. *Stress in the Public Sector, Nurses, Police, Social Workers and Teachers*, March 1988, by the Health Education Authority is still useful.

Many of the trade union guides to stress concentrate on basic alterations in the system, particularly those in non-educational areas, few advocate as I do here a more positive role for supervision. Some unions have established stress helplines for despairing remedies, but I have always felt that this encourages the individual pathology approach. Countless management orientated works focus on coping, curing and counselling the victims of stress, so that behaviour is modified rather than the system. I obviously have asked for a more balanced approach in youth and community work which has as its starting point the use of self and others, recognising the dialectical interaction between an individual and their experiences, contexts and relationships. In this sense some of the more sympathetic works on supervision need to be considered again. A good starting point is the still relevant document produced by the National Youth Bureau in 1982: *Who Takes the Strain, the Choices for Staff Support*, by Warren Feek.

## Section Six

## Violence and aggression at work

For too long violence has been considered 'just part of the job'.

Violence at work should not be tolerated and employers have an urgent duty to ensure that youth and community workers are protected as much as possible against it and that they are properly supported when they are subject to violence, or upsetting aggression. A specialist publication has been produced for youth and community workers to look at this subject in greater depth: *Managing Aggression and Violence at Work, a Model for Youth and Community Centres of Legal Compliance, Safe Working Practices and Good Personal Safety Habits for Staff*, More, W, and Nicholls, D, Pepar Publications, 1997, The Gatehouse, 112 Park Hill Road, Harborne, Birmingham, B17 9HD.

Youth and community workers are not just vulnerable by virtue of working in potentially violent situations, we are one of the few groups of workers actually employed to manage conflict and defuse aggressive or potentially violent behaviours. As a consequence of this there can be an unhelpful stigma if a worker is subject to violence. There is also sometimes a need to soak up aggression more than other groups of professionals. Equally our understanding of distressed young people and adults often leads us, as victims, to have a level of 'sympathy' with the perpetrator. No worker however, should tolerate violence and aggression towards them at work without a level of preparation and training.

It is the employer's responsibility to ensure that you are working in a safe working environment and to ensure that potential risks at work are regularly assessed. The employer's contractual obligations have been spelt out in the third section of this Pocket Guide. Violence and aggression are forms of hazards so they should be dealt with in the same way as any other risk at work. In looking at this issue it is perhaps appropriate to describe what the union can do which, in form, is no different from what it can do on any other aspect of health and safety.

## Available support

The kind of support CYWU can offer.

### Branch
The local branch is the first point of contact. If there are mounting violence and aggression problems, the branch can report this to the authority and request that a strategy to reduce risks, protect workers, train workers and issue clear guidelines is undertaken.

### Casework advice
In each branch there is a caseworker responsible at the first level for dealing with an issue. Let us say you have been persistently threatened by an eighteen year old with a record of violent and armed aggression. The caseworker has a right to meet the employer and consult with relevant health and safety experts.

### Casework representation
As well as advising you, the caseworker can represent your interests with the employers. In this instance, the caseworker would probably want the local authority to put a banning order on the young person so that they could not come onto your workplace site. Local authorities can do this in

education department premises, but it is more difficult in the voluntary sector.

### Legal support

Either you or your caseworker may seek legal advice. This can been done in two ways. For instant advice the union can arrange for you a half hour interview with a local solicitor affiliated to the Union Law Scheme. To get advice leading to representation your caseworker needs to contact CYWU's General Secretary who will then authorise a referral to the nearest NUT regional office. You will then be required to fill in a legal assistance form and to meet with the union solicitor.

Let us say that the young person actually assaulted you following the refusal of the local authority to ban the young person from your premises. You may have a claim of compensation against the local authority. Invariably to pursue this you would need a union solicitor. Each physical and mental injury has a price, however big or small, and an amount of compensation is identified in law and precedence as being associated with that kind of incident.

### Health and safety risk assessment

You may require the union to undertake a risk assessment for you at work so that you can present a fuller case to management. In this case the local health and safety representative should have the expertise. If they have not, the branch should be able to access through the CYWU region. You may also require the union to get the employer to undertake a risk assessment.

### Negotiating training

The branch may assist you by giving you application forms for health and safety training courses or by negotiating work-time training for you by the employer so that you are aware of your responsibilities and your employers requirements of you. The whole area of predicting and preventing violence at work is a specialist skill that requires training and awareness. More controversially training in basic restraint techniques though initially about as far from youth and community work values as the Stock Exchange, is an aspect that cannot be ignored for much longer.

## Other CYWU support measures

The union can respond to your branch requests either for training specific to your branch needs or providing guidance papers or developing a campaign out of your concerns. If you are hospitalised as a result of an injury at work, remember that the union can obtain funds for you if you are in hospital for more than five days.

If your property is damaged at work, or your car or other vehicle is damaged, the union can provide insurance cover of a limited amount.

## Points for consideration

An assailant is like a dangerous machine at work. S/he can cause you injury. It is your employer's legal duty of care towards you to ensure that you work in a safe and healthy environment. Therefore, within reason, the employer has a legal responsibility to prevent violence against you at work and to take reasonable steps to prevent any recurrence of the violence should it occur. They should repair the machine and your working relationship to it.

The trouble is of course, we do not work with machines, we work with young people and adults, often those most brutalised by poverty and social decay and destitution, or vulnerable to drugs and substance abuse and behaviour altering situations. Nevertheless you should not be expected to put up with murder, assault, threats and intimidation, or damage to your property.

## Post trauma

Being subject to threats, aggression and violence at work can be exceptionally traumatic, so it is always important that the employer recognises that the violent event is just the start of what could be a protracted problem. The hurt individual will want to be 100% sure that the threat of future violence against them has been removed from their workplace and that if they are feeling mentally or physically unsteady they will be supported slowly back into their duties rather than thrown in at the deep end. Careful, sometimes daily supervision of the worker may be needed.

## Violence to property

One of the most frequent ways in which workers experience violence is indirectly through attacks on their property. You as an individual are being got at through your possessions. Make yourself aware of what is covered by your employer's insurance. The answer regrettably in most cases these days is – relatively little. This is an important area for union negotiation and once you have found out to what extent the employer's public liability insurance, their professional indemnity insurances and employer's liability insurances go, you should seek to negotiate more comprehensive cover.

## Institutional codes

Your personal position will be significantly strengthened at work if your adult or youth groups have agreed themselves, through a process of group discussion, an acceptable code of practice in relation to acceptable standards of behaviour.

## Other legal and insurance matters

If you are attacked in any way at work you have the
option of involving the police and are strongly advised in
each case to report it to the police as well as the employer.
If you decide to prosecute the offending individual you
will need to take the action individually but ensure that
your employer fully supports you – after all you were
working for them at the time. Councils are often unable
because of restrictions made on employed lawyers by the
Law Society to offer direct legal representation for their
employees. However, if the LEA decides to prosecute under
Section 40 of the Local Government Act 1980
(Miscellaneous Provisions) it would meet the legal costs. So
it is always advisable to push the employer as far as
possible to take your case on. At the very least you should
ensure that the local authority will provide some free legal
advice to you where the CPS decides themselves not to
prosecute.

If you prosecute it is advisable also to take civil action
to recover compensation.

If the attack on you led to serious injuries beyond cuts,
sprains or bruises you may be able to obtain
compensation from the Criminal Injuries Compensation
Board. You can obtain application forms from the board
at 10–12 Russell Square, London, WC1B 5EN.

It is vital that the authority has taken out personal
accident and assault insurance policies to cover all of their
staff. Your union branch should check this policy; a good
one will also include cover for members of an employee's
household who is injured in an attack at the home of the
employee, provided that the attack arises from the
employment position. There will be limits to the liability
in the policies, so you should check them.

## Model statement on violence at work

CYWU and ........................................ Council agree the following policy in relation to violence and aggression at work.

- The council is committed to taking all measures necessary to combat and minimise threats of violence to its staff.

- The council will fully support staff who have been assaulted or suffered verbal or other forms of abuse and threat. Such incidences will be seen as an adverse reflection on the perpetrator and not the victim.

- The council will designate an officer to fully investigate the incident and another officer to support and counsel the member of staff and their line manager.

- The council will take out sufficient insurance policies each year to protect all members of staff. These policies and the cover will be reviewed annually by management and union through the JCC.

- In any event the council will provide one of its solicitor's to advise the members of staff on the best course of action to receive redress and compensation from the assailant.

- Where the council pursues the case all costs will be met.

- The council will do all in its power to prevent future recurrence of the incidence and where necessary and legally viable issue banning orders or other appropriate letters of warning to individuals abusing council staff.

- Confidential support and counselling will be offered to all abused staff and where necessary to their families, consideration will always be given to agreeing a phased return to work and a period of recuperation off work on full pay.

## Model incident report form

It is important for the health and safety of all staff that a record is maintained of an incidence of violence, threat or abuse of any sort.

This form can be completed at supervision or with the assistance of the line manager and will be kept as part of a confidential record with all other incident report forms within this employing organisation. Information from this form will not be divulged to any source without your permission.

# Model incident report form

**Personal details**

Name .....................................................................................................................

Age .......................................................................................................................

Gender ..................................................................................................................

Your usual place of work ....................................................................................

Your job title .......................................................................................................

Date of incident ..................................................................................................

Exact time of incident .........................................................................................

**Details of the person who assaulted/abused you**

Name .....................................................................................................................

Address .................................................................................................................

..............................................................................................................................

Gender ..................................................................................................................

Age .......................................................................................................................

Please detail any witnesses to this event



Where did the incident occur? ............................................................................

What exactly happened and what was the nature of the incident?




Had you received any training in the management of violence and
aggression at work?                                                                Yes ☐   No ☐

Had a risk assessment of your work place or your client group been
undertaken?                                                                          Yes ☐   No ☐

*Model incident report form – page 2*

If so when?..................................................................................................................................

By whom?...................................................................................................................................

Had you ever previously alerted management of any threats posed by
this individual?                                                                                      Yes ☐   No ☐

Was the individual previously known to you?                                       Yes ☐   No ☐

What was the nature of your injury?

How long were you off work as a result of the incident?..............................................................

Have you been on a phased return to work?                                       Yes ☐   No ☐

What in your view could be done to minimise such dangers in the future?

Type of abuse or assault

Please indicate if you think the incident contained elements of any of the following and explain
further if necessary:

    Attack on property

    Racial assault/abuse

    Sexual assault

    Physical assault

    Verbal/physical abuse/racist/sexist language

## Further reading

Without doubt the most readable and useful publication on this area for youth and community workers is More, W and Nicholls, D, *Managing Aggression and Violence, A Model for Youth and Community Centres of Legal Compliance, Safe Working Practices and Good Personal Safety Habits for Staff*, 1997, and More, W and Howell, A, *Handling Violence and Aggression in Education: A Personal Guide*, 1996, both Pepar Publications, The Gatehouse, 112 Park Hill Road, Harborne, Birmingham, B17 9HD. An Australian youth work lecturer Vaughan Bowie has also written on the subject and collected a huge database of related materials: Bowie, V, *Coping with Violence, A Guide for the Human Services*, Karlbuni Press, Sydney, 1989.

## General observations

In 1993 CYWU passed a comprehensive policy which instructed members not to drive buses with crew seats, or without seatbelts and to always ensure rigorous testing of buses prior to journeys, and to always drive longer journeys accompanied by qualified co-drivers. Following recent appaling tragedies minibus safety has been a matter of considerable political and social debate and new legislation. The youth service and education generally have heightened their awareness of the issues. Transport managers have begun to work closely with youth service drivers. Youth workers have been undergoing new minibus driving tests, although tests without comprehensive training are not a great deal of help. ROSPA has produced a model 90 minute minibus driver test. Some authorities have ensured that all of their buses are fully equipped with belts and safety equipment. The London Borough of Islington for example has developed some useful practice.

While there have been many positive developments, the shrinking resource background has not necessarily improved the general safety picture and voluntary sector organisations especially, have not always had new money to improve their vehicles. It remains possible for volunteer drivers to drive buses for between 8 and 17 passengers without having taken a test. Indeed, most buses used within youth and community work are not part of managed fleets. This puts additional pressures on the drivers – full and part-time youth workers. It is for this reason that I have provided what may seem, to veteran drivers, a very basic guide. It is not my intention to duplicate the very expert advice promoted by ROSPA or the Community Transport Association.

This section should be read in conjunction with the advice of the transport experts and the focus here is not initially for the experienced driver. I have in mind the recently qualified JNC full or part-time worker entering their first post in a voluntary sector organisation with its own dilapidated minibus which is not part of a fleet. No matter how frequently you will be called upon to drive a minibus, the full range of considerations mentioned here and the expert advice referred to should be taken account of. Youth and community workers are not drivers by profession. They drive as a small but significant part of their professional duties. In this sense they should rely on the expert support of transport professionals in ensuring that what they do and what they drive is completely safe and well understood. Youth workers are not mechanics or long distance lorry drivers. It is inevitable that check-lists and prompts are a feature of the format of this section.

Under no circumstances whatsoever just jump into a bus and drive away. Use the following check-list prior to any trip, short or long, build time into your schedule to run through it before any trip.

## Minibus safety check-list

**Do not drive a minibus anywhere unless:**

- You have been fully instructed in safe minibus driving and tested.

- You have third party, professional confirmation in writing that the minibus is fully roadworthy and safe to drive. You should have evidence of regular care and maintenance of the bus you drive and of all the necessary documents. In addition you should be able to undertake a routine vehicle inspection: tyres, battery, lights etc..

- You have passed the locally recognised test within the service. (Please note that a European Directive that came into force on July 1st 1996, requires new drivers to take a test before they can drive a minibus with more than eight passengers, other than one used purely for social or voluntary purposes. (Details of the types of licences you require are available from Community Transport Association, High Bank, Halton Street, Hyde, Cheshire, SK14 2NY. Send a stamped addressed A4 envelope. See also further reading section below).

- You are aware of all of the lines of insurance, health and safety and accident reporting responsibilities and the contact points for each.

- You are sure that the minibus you will be driving is completely safe and checked with all of its documentation up-to-date. If in doubt refuse to drive and refer to your union for health and safety advice. You should have available to you a comprehensive physical check-list of all aspects of the vehicle. How do you check tyres, how do you check brakes?

- The minibus has forward facing passenger seats throughout and is fitted with three-point diagonal seat belts throughout. (Please note that all road safety experts and organisations advise that buses with sideways facing crew seats are only intended as utility vehicles designed for small scale ferrying of people around for example building sites. They are not intended for motorway and long distance driving. Do not use them. In addition, lap only seat belts are not recommended).

- You have a list of the names and addresses of all those you are transporting on your person during the trip and left at work with another person. Such details should normally include parental consent forms etc..

- The minibus is appropriate for the user group. Do not wrestle wheelchairs into a vehicle with no proper lift or spacing. There should be one seat for every passenger.

- You have a route plan and have informed someone at base of this and the likely times of your arrival at points en route and your destination.

- You have been trained in accident and emergency procedures.

*Minibus safety check-list – page 2*

- All luggage is firmly secured, preferably on a roof rack, if not in a trailer.

- There is ample space between the rear of the bus and the passengers.

- You have applied a risk assessment to the whole journey from preparation to conclusion. The structural model of the risk assessment is as outlined in the earlier section above, the following ingredients should be considered in relation to minibus driving.
  - identify hazards
  - decide who might be harmed and how
  - evaluate degree of risk and adequacy of existing precautions and identify further action to control the risk
  - record findings in writing
  - review assessment
  - report risks
  (Note the Community Transport Association can provide guidance on risk assessment)

- Driving times and hours are reasonable and there is a qualified co-driver for journeys of a total of six hours.

- You have checked that the minibus complies with the statutory requirement to have an approved fire extinguisher and a first aid kit on board at all times.

- You have access to a mobile phone on the journey.

- You are accompanied on the journey by another adult colleague who has ideally passed the minibus test.

The check-list above is minimal and basic. In general, a driver should ensure that a minibus and the trip being undertaken are subjected to a full risk assessment. What will be potential loopholes, hazards, problems? A driver will also need to be aware that they have rights under health and safety to refuse to drive unless they are fully confident that all health and safety guidelines are met. Some of its key contents need further elaboration as detailed in Section 3 above and drivers will need to be aware of local practice. Drivers taking groups in a minibus abroad will also need to ensure that they are fully trained in all of the regulations relating to travel abroad.

## Seatbelts

A major problem in minibuses is that seatbelts were previously usually provided on the front driver's and one passenger seat. There is now a requirement to provide them throughout and really for use with young people one full seat belt per seat is necessary. The belts should conform to European standards (the belt should display an "e" or "E" mark or a British Standard kitemark BS3254 1960 or BS3254, part 1, 1988. The belts need to be firmly anchored to the stronger chassis parts. To provide seatbelts to existing front facing minibus seats is a costly business. Also, where seat belts are added to some buses it is not possible to add them to safe anchorage points, but only to link them to the seat frame itself. This is not the safest option. Another problem is that often lap top only seat belts are fitted rather than the familiar car type seat belts which anchor at three points and cross the lap and shoulder. An antiquated aspect of legislation permits three children under 14 to sit on each double seat.

This risky background of manufacturers' practice and legislation at best means that youth and community workers will have to adopt a strategic approach with their employing organisations to improving minibus safety. However, safety can never cost too much and unless drivers take a firmer stance on refusing to drive unsafe vehicles they will continue to jeopardise the lives of user groups. I am therefore firmly of the view that immediate action should be taken to ensure that no youth and community worker is expected to drive a minibus unless:

- Diagonal seatbelts are anchored to the frame (proper anchorage points) of the vehicle rather than the seats. Any modifications to fit diagonal seat belts should fully comply with the 'Construction and Use Regulations'.
- One seat with a diagonal seatbelt is available for every passenger.
- Any lap top belted seat in the front is left vacant.
- Young people can be transported with one seat each.

## The vehicle

In addition to the routine safety checks and roadworthiness criteria for which you should have written evidence from the log book and service report, a driver should bear in mind the in-built inappropriateness for carrying passengers of many minibus designs. This is because many are based on commercial van frames designed for carrying loads rather than people and their luggage. One feature of this is the closeness of back seats to the rear doors. Drivers should really ensure that there is an adequate space between rear doors and the first passengers, even if this means leaving some rear seats empty. This is known as a crush area and would act as a buffer if there was an intrusion in an accident from the rear. You could consider using the rear seats as luggage carriers, or removing the rear seats completely and using this space for luggage.

On youth and community trips you will usually have luggage. Ensure that this does not obstruct passengers and that it is firmly secured. Unsecured luggage in a crash or emergency braking become lethal flying weapons. Most road traffic experts recommend separate luggage compartments in the form of a roof rack, luggage bins, or a trailer.

Seats in the minibus should all face forward. There should be ample room for entry and exit, especially bearing emergencies in mind. Gangways to all doors should be clear and not obstructed by luggage or passengers. If it takes you five minutes to haul a wheelchair on to a vehicle without a lift, how long will it take you to haul it off in a crash.

The Royal Society for the Prevention of Accidents is advocating that future safeguards on minibuses should include automatic fuel tank and electrical cut off devices and anti-burst fuel tanks. It is important to enquire about manufacturers' policies in relation to these when purchasing a new bus.

## Responsible enough to drive?

Youth and community workers are educators not chauffeurs or mechanics. Driving is a means to an educational end, and very often the discussion with the young people or adults in the minibus is an integral part of the group building and educational experience. Very often the driver's real work begins when the journey has ended or in the preparation for it. Youth and community workers are always on duty on a trip. This duty is in fact circumscribed legally.

As the Children's Legal Centre's document *Working with Young People: Legal Responsibility and Liability* indicates "Anyone employed to work with children or young people is under a legal 'duty of care'. This duty has been defined in case-law as acting as a careful parent would. A working definition of this has been 'The body/organisation/authority/employer should take such precautions for the safety of the children in their care as would any reasonable caring parent'." If you do not do this and as a result of your negligence an injury, accident or loss takes place, you and your employer could be taken to court in civil law. It is essential not only to be fully competent and proficient to drive, but also to be fully aware of the indemnity and liability cover applying to your activities.

A driver should be trained and tested to drive. Each trip should be properly assessed in terms of the likely total workload associated with it. The chances are that the youth and community worker will be up early to panic about attendances, equipment packed, safety of the bus, payments and so on. They will then deal with the high spirits of the passengers and the worries of parents. They will then have a long drive and they will then risk being up late on the first night for the young people away from home. As an integral part of the risk assessment it is therefore vital that the total working time in relation to the trip is taken into account and also the likely times when driving will occur. Obviously there will be times during a day when a driver is less alert for driving. All accident research demonstrates that a driver who has been awake for more than seven hours and then drives for more than two hours is at far greater risk of crashing. Similarly, if drivers drive into 'normal sleep times' they increase their risk of accident. If a journey itself will take more than six hours a second driver should be provided.

ROSPA itself is very clear that drivers have a rest and refreshment break at least every six hours. They further recommend that non-professional drivers should not be allowed to drive for more than a total of six hours on a journey even with regular breaks.

If your task is to drive with a co-driver then you will need to concentrate on this. Passengers with special needs, or requiring special attention and supervision should be provided with escorts. Make sure you are appropriately staffed and that problems unrelated to the task of driving are capable of being solved by other colleagues.

## Licensing arrangements

Regulations applying to school teachers should be considered applicable in this sector too. Important points within new arrangements worthy of note are as follows:

- teachers cannot be required to drive minibuses;
- teachers must satisfy a license, and where necessary, the permit requirements for minibus driving;
- teachers should receive training in driving minibuses;
- teachers are personally responsible for the roadworthiness of vehicles they drive;
- certain pre-drive safety checks must be undertaken.

Where teachers volunteer to drive a minibus the teacher must by law;

- hold a valid full driving licence to do so;
- be at least 21 years old;
- be insured to drive the vehicle in question; and
- have held a full driving licence for at least one year when driving a minibus under a Section 19 permit.

It should be noted also that drivers who passed their car driving test after January 1st 1997 no longer automatically gain additional Category D1 entitlement to drive minibuses. In these circumstances the individual is required to pass an additional Category D1 test to drive minibuses other than in certain circumstances. See Factsheet INF28, *Driving a Minibus in Great Britain*, published by the Department of Transport and available from the DVLA on 01792 772151.

## Further reading

The Community Transport Association. High Bank, Halton Street, Hyde, Cheshire, SK14 2NY, produces an excellent and comprehensive publications list including: *Your Minibus – Is it Legal?, Starting Up, Identifying Hazards, Minibus Safety Charter, Driver Assessment and Training Pack, Code of Practice etc.. The Operation of Minibuses in the Voluntary Sector.*

*Community Action Driver Information Pack*, Nottingham Community Action, Portland Building, University Park Nottingham, 1990. Free.

*Youth Clubs and the Law: Applying for a Minibus Permit.* In Youth Clubs, November 1989, no 55, page 45.

Dring, Tony, *An Introduction to Basic Minibus Driving.* Also, *Essential Minibus Driving.* Royal Society for the Prevention of Accidents, Cannon House, The Priory, Queensway, Birmingham, West Midlands, B4 6BS. ROSPA provide a range of information sheets including: *Minibus Fact Sheet: Seatbelts, Inspection, Safety, Seatbelts for Minibuses and Coaches*, and a minibus driving test. Contact their driver service on 0121 706 8121.

*The Resource Project – Transport.* Marinie Easton and Christine McElligott, Islington Play and Youth Service, January 1996.

Department of Transport, 2/06 Great Minster House, 76 Marsham Street, London, SW1P 4DR. DVLA, *Important News*

*for Drivers of Minibuses*, Fact Sheet, March 1994. Free.
*Minimum Test Vehicles*, Fact Sheet, March 1994, *Towing
Trailers*, Fact Sheet, April 1994. *Advice to Users and Operators
of Minibuses and Coaches Carrying Children*, VSE 1/96). Note
many of these documents are available on the Internet at
(http://www.open.gov.uk.home.htm).

National Union of Teachers, *Safety on School Journeys*.
Leaflet, Free, NUT, Hamilton House, Mabledon Place,
London, WC1H 9BD.

*Minibus Guidelines: Guidance for Youth Workers in Using
Minibuses*. Newham Youth Services, 1995.

*Minibuses: A Code of Practice for Vehicles Operated Under
Permit from the Council*. London Borough of Croydon
Education Department, 1993.

# Harassment and bullying at work

Two of the most frequent stress inducers leading to illness in youth and community casework files are harassment, usually of a sexual nature, and bullying. In this section I provide some details of the excellent resources that exist quite widely now and a model procedure on harassment and comments and procedure on bullying. Following some recent important court cases there is absolutely no doubt whatsoever that the employer is responsible for the harm caused to an employee by the torment of another employee within the same employing organisation.

# Model policy on sexual harassment

**What do we mean by sexual harassment?**
*(This first section, taken from national CYWU policy may be usefully reiterated in terms of the employing organisation or the local branch.)*

Unwanted and sometimes persistent sexual comment, looks, suggestions or physical contact. The woman finds these advances objectionable, threatening, patronising, humiliating, and is caused extreme discomfort in her job as a result.

It is recognised that men can be harassed too, with gay men in particular sharing with lesbian women the indignity of heterosexist harassment. But this organisation believes that power relations within a workplace mean that in the majority of situations women are harassed by men.

Sexual harassment is not the private problem of individual women, but is a product of the deep sexist divisions in society, it is a public matter. Since trade unions exist as expressions of unity and equality of all their members it follows that discrimination, oppression and harassment of all their members should be opposed. CYWU therefore has no hesitation in declaring that sexism in general, and sexual harassment in particular, are matters of Trade Union concern and require the collective support and action of all its members.

**What are its effects?**
- A woman may be forced to resign from her job to escape harassment.
- She may be refused promotion for not complying with the harasser's demands.
- Her performance as a worker will be affected by the distractions and tensions of the situation.
- The stress caused by sexual harassment as well as affecting the complainant's personal health also creates a health and safety problem at work.
- Sexual harassment reinforces and perpetuates women workers' second-class status. It looms always as a reminder to women that they are seen not as serious and equal colleagues, but as convenient sources of decoration and titillation.
- Sexual harassment is not just a part of everyday dialogue and interaction to be enjoyed or ignored, rather it is a clear demonstration of power over women which must not be endured.

**What should we be doing about it?**
Educating and campaigning: members must be aware of what sexual harassment is, what damage it can do, and how it can be dealt with. Many women hesitate to report harassment:
- They don't think they will be believed.
- They think their working relationships will become worse.
- They think they will be punished by transfer or by refused promotions, etc..
- They will be seen as lacking in humour by female as well as male colleagues.
- They fear they may be accused of inviting the advance, and begin to feel guilty.
- They fear the publicity the confrontation may cause.
- They are unsure about what exactly they should understand as harassment and what is acceptable everyday behaviour amongst colleagues.

Education could begin with organised meetings of women members.

Questionnaires on harassment could be circulated nationally and used for discussion.

Campaigning through posters, leaflets, alternative calendars.

## The effects of sexual harassment

- A woman may be forced to resign from her job to escape harassment.
- She may he refused promotion for not complying with the harasser's demands.
- Her performance as a worker will be affected by the distractions and tensions of the situation.
- The stress caused by sexual harassment as well as affecting the complainant's personal health also creates a health and safety problem at work.
- Sexual harassment reinforces and perpetuates women workers' second-class status. It looms always as a reminder to women that they are seen not as serious and equal colleagues, but as convenient sources of decoration and titillation.
- Sexual harassment is not just a part of everyday dialogue and interaction to be enjoyed or ignored, rather it is a clear demonstration of power over women which must not be endured.

## What should we be doing about it?

Educating and campaigning: members must be aware of what sexual harassment is, what damage it can do, and how it can be dealt with. Many women hesitate to report harassment because:

- They don't think they will be believed.
- They think their working relationships will become worse.
- They think they will be punished by transfer or by refused promotions, etc..
- They will be seen as lacking in humour by female as well as male colleagues.

- They fear they may be accused of inviting the advance, and begin to feel guilty.
- They fear the publicity the confrontation may cause.
- They are unsure about what exactly they should understand as harassment and what is acceptable everyday behaviour amongst colleagues.

Education could begin with organised meetings of women members.

Questionnaires on harassment could be circulated nationally and used for discussion.

Campaigning through posters, leaflets, alternative calendars.

A sexual harassment clause must be included in the equal opportunities policy of the employer, stating that neither union nor management will tolerate sexual harassment at work. There will be a training programme on sexual harassment for officers and workers and, because there are different objectives, this training will be separate for men and women; and that there are joint union/management procedures for dealing with cases of sexual harassment.

## Grievance procedure

The union and the employers recognise the problem of sexual harassment at work, and are committed to ending it. Sexual harassment shall be defined as:

1. Unnecessary touching or unwanted physical contact.
2. Suggestive remarks, hints and innuendoes or other verbal abuse.
3. Leering at a person's body.
4. Compromising invitations.
5. Demands for sexual favours.

6. Physical assault.
7. Displaying pin-ups.
8. Any other behaviour identified as sexually harassing.

Grievance under the above clause will be handled with all possible speed and confidentiality. In settling the grievance every effort will be made to discipline and relocate the harasser, not the woman.

The procedure should cover the following stages:

**Stage 1.**

a. A trade union member who is being sexually harassed should seek advice from the CYWU including its women's caucus where appropriate, during the course of proceedings.

b. If a woman does not feel able to ask the harasser to stop, then the trade union representative must carry out the task and make clear that the behaviour is unacceptable.

**Stage 2.**

a. If the complainant is not satisfied with the response to this initial approach or the harassment still continues, the complainant together with the trade union representative should take up the matter with the youth officer (or management committee members). An oral reply to this dialogue must be received within two working days at the most.

b. If the harassment still continues, meetings are to be organised between the representatives in a further attempt to resolve the matter before proceeding. This is to be done within one working week. The union should indicate at this stage that it may recommend that the member withdraw from the work vicinity of the harasser as a health and safety hazard.

### Stage 3.

a. The grievance should be submitted in writing to the appropriate LEA officer (management committee) with the worker and trade union representative retaining copies. The officer must then inform the person identified as the harasser that the matter is being taken further, and advise him that he is entitled to seek trade union advice.

b. Within 10 days of the officer receiving the complaint separate investigatory meetings to be called with the complainant and the accused together with their representatives.

c. As soon as possible after the meeting, two officers, one of whom should be a woman of the complainant's choice from within the authority or organisation, are to give their judgment and explain to all concerned exactly what further action is to be taken.

d. If this includes disciplinary action against the accused, disciplinary procedure shall be instigated.

e. If either party is dissatisfied with the judgment he or she has the right of appeal to an appeals panel.

### Stage 4.

a. Appeals panel, which should be not less than 50% women, to be convened within 10 working days of a written request. The judgment of the panel to be communicated in writing as soon as possible.

There will be no reference made to the above procedures on the woman's file. (The officer should decide what goes on the harasser's file, a judgment of the gravity of the offence, and the length of time that it should stay on the file).

## Sexual harassment – key elements

- A woman member with a complaint of sexual harassment can seek advice/support from the CYWU branch, local caseworker, national caseworker and women's caucus.
- If she so wishes the union will initially approach the harasser.
- There is a clear opportunity for informal settlement before anything is out in writing.
- Once in writing, the alleged harasser has the right to be informed.
- Before formal investigation, resolution can be sought at an informal meeting.
- If necessary, disciplinary action can be taken under the existing procedure.
- Both parties have the right of appeal.
- No reference to the procedure will be placed on the woman's file. What is placed on the harasser's file and the length of time it will remain depends on the gravity of the offence.

In addition it should be remembered that an act of harassment by a CYWU member is clearly against the union's code of conduct and should be treated as a disciplinary matter within the union where appropriate.

Many of the points, principles and stages outlined above will relate to the question of racial harassment as well. Harassment after all is behaviour which is unwanted, unreciprocated and offensive to another.

## Bullying

This is a form of individual harassment and victimisation. It is defined most commonly as: "persistent, offensive, abusive, intimidating, malicious or insulting behaviour, abuse of power or unfair penal sanctions, which makes the recipient feel upset, threatened, humiliated or vulnerable, which undermines their self confidence and which may cause them to suffer stress." Bullying as in other forms of harassment involves an abuse of authority and power relationships, but not always. Whatever the pathology of the individual bully, the effect of their activities is the concern here.

A bully may use terror tactics, naked aggression, shouting, abuse, threats and obscenities. They may seek to humiliate, ridicule and belittle their targets' efforts and achievements, sometimes in front of others. In youth and community work a hallmark of the bullying manager is the excessive use of close supervision, to drown routine tasks in comment and criticism, being excessively critical about minor things. Sometimes they will seek to get credit for the work of the person they are bullying, but blame them behind the scenes or openly when things go wrong. They may seek to overrule a person's authority, they may strip a victim's job of its best elements and reduce it to the routine. They may frequently change the work responsibilities or ground rules without fully conveying the information about the changes involved. They may withhold essential information. They may ostracise their victim, dealing with them only through a third party. They may spread malicious rumours about an individual quoting part of a story or a written correspondence over the phone to third parties without their victim being able to explain their point. Bullies enjoy controlling things like leave, timesheets, requests for training and blocking promotion and job satisfaction opportunities.

There is no legislation which specifically outlaws this kind of unacceptable behaviour, but its negative effects on a victim's health needs to be placed within the framework of health and safety legislation and codes of practice outlined elsewhere in this publication. There is of course legislation which makes certain forms of discrimination unlawful. Bullying could be construed as discriminatory where it involved demonstrable and significant elements of sexual or racial discrimination or harassment. In such cases the terms of the Sex Discrimination Act 1975 and the Race Relations Act 1976 will be relevant. Additional guidance is given to employers through various codes of practice on these matters:

- Code of Practice: *Equal Opportunities Policies, Procedures and Practices in Employment* (Equal Opportunities Commission 1985).
- *Code of Practice on Sexual Harassment* (European Community 1991)
- *Race Relations Code of Practice* (Commission for Racial Equality 1984).
- Code of Practice: *Equality of Opportunity in Employment* (Employment Equality Agency 1983).

It is obviously going to be advantageous to have the spirit of these codes of practice incorporated in an anti-bullying policy and procedure built into every workplace.

## Further reading

CYWU was one of the first unions to put a sexual harassment policy into its formal policy document. This forms much of the section above. Many unions have good materials on these subjects. MSF's work on *Bullying* is particularly good and in many ways led to the trade union recognition of the subject. UNISON and the NUT have produced good materials on harassment such as: *Sexual*

*Harassment is an Abuse of Power*, UNISON (0181 854 2244 exts 252/226) *Bullying at Work* from UNISON is also useful. *Bullying at Work: How to Confront and Overcome it*, by Andrea Adams, Virago Press, is a substantial book length publication on the issue for those wanting to go into more depth. *Tackling Harassment at Work*, by LRD is essential reading also.

On the question of racial harassment and abuse there is now a wealth of support material also, much of which is available in CYWU national office.

In addition the following books are worth getting: Adams, Andrea. *Bullying at Work*, Virago. *Bullying at Work – Combatting Offensive Behaviour in the Workplace* – a training package including two videos, available from the BBC, Woodlands, 80 Wood Lane, London, W12 0TT. Tel: 0181 576 2361.

*Tackling Harassment at Work* from the Labour Research Department 78 Blackfriars Road, London, SE1 8HF is also good. There is a National Harassment Network at the University of Central Lancashire, Preston, PR1 2HE. Tel: 01772 893 398. The Industrial Society also provides advice on bullying 0171 262 2401. The British Association for Counselling can provide support information too, 37a Sheep Street, Rugby, Warwickshire, CV21 3BX. Tel: 01738 78328/9.

# Education outdoors, and on trips and visits

## A model policy statement on educational visits, exchanges and trips

Parishire County Council recognises that education trips, exchanges and journeys are an integral part of the youth service curriculum and that the highest standards of educational experience and health and safety must be achieved. In drawing up these guidelines the intention has been to support this work rather than constrain it, while creating an environment of care that will protect the participants, staff and employing organisation to maximise educational opportunities. It is intended that the county council through adherence to the guidelines will exercise all reasonable care for its employees and that its staff will in turn be enabled to exercise their duty of care to each other and the participants in their charge.

The procedures contained in part 1 of this document must be followed in planning, arranging and undertaking visits and journeys. Disciplinary action could be the consequence of any failure to comply.

The requirements made in this document apply to all those, regardless of age, participating in excursions and trips organised by the county council. The requirements must be known by, and adhered to equally by, full-time, part-time and volunteer staff.

It is recommended that these county council procedures and requirements are made known to all youth and community service users and particularly to those participants seeking involvement in one of the specified activities.

The county council will provide details separately of the following legislation and regulations which govern the organisation of excursions:

- Health and Safety at Work Act (1974) – Safe practice at work.
- Management of Health and Safety at Work Regulations (1992) – Risk Assessment.
- Activity Centres (Young Persons Safety) Regulations (1996) – Centre safety, instructor qualifications.
- Education Reform Act (1988) – Acceptable excursions and charging etc..
- Occupiers Liability Act (1957).
- First Aid at Work Act (1982) – Accident reporting and first aid.
- Transport Act (1985) – Safety of minibuses and coaches including driver hours etc..
- Minibus Regulations (1987) – Minibus permits, charging for voluntary groups.
- Public Service Vehicles (Carrying Capacity) Regulations (1996) – Number of passengers, seat belts etc..
- Package Travel, Holiday and Tour Regulations (1992) – Insurance cover and emergency support.

## Approval

The approval of the area youth service manager/principal youth officers must be obtained prior to undertaking any visit or journey which:

- involves an absence for one night or more
- involves travel by air or sea
- involves a hazardous activity (as defined later in this document).

Approval must be obtained in writing and must be requested on the appropriate form setting out the full details of the activity.

Youth workers responsible for units may themselves authorise visits or journeys which do not come under the categories outlined above. Reference will be made in any event to these guidelines and other codes of good practice in operation for such activities. Guidance must be sought from the appropriate body as to the nature of the activities and the level of risk involved. The worker responsible will undertake a risk assessment of the activity bearing in mind the age of the group concerned, their experience and the nature of the supervision required. Full details of the activity and the itinerary will be left with a responsible adult.

In seeking approval the following information will be given:

- educational objectives of the trip
- purpose
- cost to participants
- cost to service
- nature of activities to be undertaken
- details of risk assessments undertaken

- date
- form of transport
- number of participants
- numbers of staff
- qualifications of staff
- number of drivers attending with minibus test pass or licence
- experience of each staff member of such visits
- details of accommodation to be used
- detailed itinerary
- name and address of any other organisations involved
- dates
- insurance arrangements

Approval should be sought at least four weeks prior to the proposed event. No financial or other commitments shall be made prior to the written approval being received. It will be the responsibility of the area youth service manager to verify any special requirements from the local authority solicitors.

### Staffing ratios

Four categories of excursion are identified.

### a) International visits

A minimum of two qualified youth workers will be required to undertake an international visit and in any event the ratio of staff to young people should be 1:8. Young people must always be accompanied by an adult of the same sex. Supervision arrangements for visits involving home to home exchanges will require case by case description.

(Please note it is the responsibility of the worker in charge to ensure that all appropriate immunisation and vaccination procedures have been carried out. A pamphlet on practice is available from the Department of Health *Notice to Travellers – Health Protection* SA 35. Individual forms should be submitted to the county medical officer.)

b) **Visits away from home over one night**
A minimum of two workers shall accompany a group leaving home for more than one night and both male and female workers should accompany mixed groups. At least one staff member should have previous experience of such visits. In any event the ratio shall be 1:10.

c) **A visit of one day's duration**
A minimum of two workers shall accompany each day visit and in any event a staffing ratio of 1:10 shall be established. At least one member of staff should have previous experience of such visits. Both male and female workers should accompany mixed groups.

d) **Visits of less than a day's duration**
A ratio of at least 1:10 is required for all visits to to other local authority establishments within or outside the local authority area and for any period of time less than a full day. Both male and female staff shall accompany mixed groups.

On all visits the number of participants, their general level of experience, maturity and other special needs will be assessed alongside the capabilities, experience, qualifications and maturity of staff. Full account of the hazardous activities must be taken including any recommendations from the relevant governing body.

Where a worker is also the driver of the minibus or other vehicle used to transport a group, that worker shall be excluded from the staffing ratio calculation.

On all group excursions a worker in charge and a deputy shall be appointed.

In addition staffing ratios for other specific activities are detailed below.

### Parents/guardians

For young people under the age of 18 advance notice of the detailed arrangements for approved trips shall be given to all parents/guardians. This information shall detail all of the activities to be undertaken, the staffing ratios, the size of the group and the level of supervision and insurance to be provided.

Parental/guardian consent forms must be obtained for young people to participate in all activities as detailed above.

## Parental consent forms

Such is the fear within the service currently, that many workers are issuing consent forms to take a group of young people to a pool tournament in a nearby youth centre. On the principle that accidents can happen anywhere, workers in local authorities in particular are playing 'better safe than sorry'. However, there are real dilemmas about consent forms. What do you do if there are eleven young people in the minibus waiting to go out for the evening and the twelfth young person, who most desperately needs the trip, has forgotten their form? The area is unsafe at night and you know that if the young person stays behind they will be at risk or get into trouble. This may be a moral dilemma for the youth worker, but for the employer it is straightforward; in the event of an accident involving the young person without the form if they were taken on the trip – the worker would be disciplined. What activities do you actually need a consent form for – what is hazardous? Employers usually attach consent forms to a policy relating to excursions and trips in which they have a list of hazardous activities which is invariably described as non-exhaustive. Nine times out of ten the accident occurs in an activity that has not been listed. What if you have issued the consent form, you know the parents have signed it because they told you so on the phone, but they haven't sent it back? Make sure there is a rule to cover this problem and similar eventualities. The exception has a habit of forming the rule.

The point is to answer these questions at the level of policy and instruction so that there is coherent practice within the employing organisation and all staff are clear of the parameters. In the light of the growing difficulties in this area the model consent form below is perhaps more stern than those with which you are most familiar, but casework is always a hard teacher. The form also requires that you are able to be clear and transparent about the level of insurance cover for each activity. This implies of course that you need to be clear of the insurance company's requirements for supervision and protection. You should ensure that you have a copy of the relevant insurance cover documents available.

## Model parental consent form

Down Town Community Centre
78 Beech Road, Paris, Tel: 01111 11111.

Worker responsible for this activity/trip .................................................................................................

Please call if you have any queries

**To the young person handed this form**
You will not be allowed to participate in this trip unless this form has been signed by your parent/guardian and returned by ............................................ (date). This is for reasons of your own health and safety, and our concern to see that the trip is properly organised and that we all have a good time. So, remember, no form returned, no go on the trip.

If you are 18 years or over you may complete the form yourself, but it must still be returned.

**To parent/guardian**
We want young people under the age of 18 years to enjoy our activities to their full and to feel secure and protected during their participation in them. Therefore, we will not be taking any young people on trips or journeys with us unless this parental consent form has been completed and returned.

Please ensure that this form is completed and returned to the address above by
.................................................................... (date).

1.  I, the undersigned ....................................................... (name of parent/guardian or young person 18 years or over) being the parent/guardian/participant over 18 years hereby give permission for him/her to take part in .......................................................................(detail activity).

    taking place on/between ................................. and ................................. dates.

2.  I have read the information sheet regarding the activity/trip and understand what is involved. I acknowledge the need for obedience and responsible behaviour on his/her part throughout the period and the need for him/her to take special note of any safety instructions. I am satisfied that all reasonable care will be taken for the safety of those participating and that adequate staffing and other insurance and safety measures have been taken. I understand the extent and limitation of the insurance cover provided. I understand that my son/daughter/ward will not be able to participate unless this form has been returned completed by me.

    I understand also that on visits to theme parks/recreational centres etc. and resort towns, close supervision by staff will only be possible while everyone is travelling on the bus, but that young people with learning difficulties or physical disabilities will be closely supervised at all times.

    I consider my son/daughter to be medically fit to participate in the activities outlined.

*Model parental consent form – page 2*

I require that my son/daughter be excluded from the following ......................................
......................................................................................................................

I would like you to be aware of the following special needs of my son/daughter ......................
......................................................................................................................

3. Declaration

   a. I agree to ........................................................ (name) receiving emergency medical treatment, including anaesthetic, as considered necessary by the medical authorities present.

   b. The person to contact in case of emergency during this activity is:
      Name: ......................................................................................................
      Address: ..................................................................................................
      Tel: ..........................................................................................................

   c. Should none of the above be available, please contact:
      Name: ......................................................................................................
      Address: ..................................................................................................
      Tel: ..........................................................................................................

   d. The participant's family doctor is:
      Name: ......................................................................................................
      Address: ..................................................................................................
      Tel: ..........................................................................................................

   e. The participant is allergic to the following medicines/foodstuffs: ......................................
      ......................................................................................................................

   f. The participant's blood group is: .......................................................................

   g. The participant's date of birth is: ......................................................................

Signed: ..............................................................................................................

Date: ................................................................................................................

Please note that your daughter/son will not be able to participate in this activity unless all parts of the above form have been completed.

Obviously the completed form should be taken to the activity by the responsible worker and be available at all times. Ensure that staff understand all the hazards and levels of supervision available at all times prior to issuing the forms so that you can legitimately answer all parents' questions before you leave.

## Activities deemed hazardous

Hazardous activities may include the following:
- Camping
- Walking in remote areas, cliff pathways and beaches or marshland or hills
- Mountain walking
- Rock climbing
- Skiing, snow and artificial slopes
- Caving, potholing and mine exploration
- Pony trekking
- Horse riding
- Cycling
- Windsurfing
- Jet skiing
- Sailing
- Sail boarding
- Water skiing
- Canoeing
- Rowing
- Underwater swimming
- Swimming, diving in the sea, rivers, lakes
- Angling
- Orienteering
- Canal boating

- Nightwalking
- Airborne activities (including airborne transport)
- Field studies involving any of the above.

This is not an exhaustive list and it is recognised that various indoor and outdoor sports and attendance at theme parks or recreational centres can involve activities that are hazardous. It is therefore recommended that the parental consent form is issued in most circumstances and, where it is felt by the responsible worker that such a form is not required, that notice is given to the area manager/principal officer to this effect in advance of the activity. Further information can be gained from the following specialist outdoor education and PE Advisers:

.............................................

There are additional hazards at various locations where groups may go, from a part of a town noted for criminal activity, to an area of the countryside with its own particular hazards. It is expected that such factors will be taken into account in planning the trip and assessing risks.

### Risk assessment

A check-list for risk assessment in group activity outdoors or for trips is attached. It is recognised however, that the prevention of accidents in any activity cannot be fully guaranteed by regulations and qualifications, but risk can be reduced if the following are considered:

a) The quality of leadership – experience, qualification, sound judgement, familiarity with the group. Evidence of this. First aid competence. Competent in rescue and survival techniques. Physical fitness.

b) Knowledge and skill in selecting equipment and understanding of natural elements to be encountered.
c) Checks on equipment before departure.
d) Safety procedures known by all participants relating to all aspects of the trip.
e) Gauging levels of protection required, probability of accident given experience of group and nature of likely responses of young people involved.

Further details and advice on these matters can be given to you before you go on your trip

by.........................................
Tel:.........................................

The following written guidance and information can also be obtained from the Resource Centre (See further reading and resources section below).

### Specific activities

This section should be subject to careful addition, amendment and verification. Its structure is indicative and policy should err on the side of inclusion rather than exclusion so that basic details can be quickly gained on each activity.

Different activities may require different staffing ratios and confirmation that the leaders/instructors involved have specified qualifications. The following requirements of staffing ratios are given, together with qualifications that must be verified and specified in the request to the area youth service manager/principal youth officer for approval. In each case complete 'further advice' with information obtained from a local expert.

## Rock climbing
*Staffing ratio* – 1:6 when top roping; 1:2 when lead climbing.
*Qualifications* – None. Extensive experience and knowledge of
  terrain required.
*Safety* – Helmets in all situations. Complete equipment check.
*Further advice:*


## Winter climbing
*Staffing ratio* – 1:6 when in training situations; 1:2 when lead
  climbing.
*Qualifications* – None.
*Safety* – Extremely experienced leadership required. Superior
  standard clothing and equipment.
*Further advice:*


## Skiing
*Staffing ratio* – Absolute maximum 1:12.
*Qualifications* – Ski course organiser's certificate/Artificial ski
  slope instructor/British Association of Ski Instructors.
*Safety* – Under qualified instruction. Protective clothing and
  equipment suitable to conditions.
*Further advice:*


## Camping
*Staffing ratio* – 1:8 in all circumstances.
*Qualifications* – None.
*Safety* – Supervised cooking. Correct equipment. Stoves
  outside tents. Care in changing fuel supplies, storage and

transit of fuel. Age limits imposed within group for use of each item of equipment. Only use equipment following instruction. Advice on fire procedures. First aider available at all times. Mobile phones available at all times.
*Further advice:*

## Mountain walking
*Staffing ratio* – 1:8 summer; 1:6 winter.
*Qualifications* – Summer: Mountain Walking Leader Assessment Scheme. Winter: Winter Mountain Walking Leader Assessment Scheme.
*Safety* – Refer to appropriate national governing bodies.
*Further advice:*

## Pony trekking/horse riding
*Staffing ratio* – 1:8.
*Qualifications* – British Horse Society Instructor.
*Safety* – Protective headwear conforming to BSI. 6473 or BSI.4472 standard. Heeled footwear must be worn. Pre-tuition necessary. Match animals to riders.
*Further advice:*

## Caving/potholing and mine exploration
*Staffing ratio* – 1:10.
*Qualifications* – Local Cave Leader Assessment/Cave Instructor's Certificate.
*Safety* – Consult National Caving Association standards. Detailed risk assessment required for each particular visit.

*Further advice:*

## Orienteering

*Staffing ratio* – 1:10.

*Qualifications* – British Orienteering Federation/teacher's certificate, coaching certificate etc..

*Safety* – Protective clothing and footwear. Defined boundaries. Accounting procedures for all participants.

*Further advice:*

## Angling

*Staffing ratio* – 1:10.

*Qualifications* – None.

*Safety* – Protective footwear and clothing. Accompanying staff proficient in expired air resuscitation and life saving award. Check tides, mud banks. Where using boats seek clarification from river or water authority or coastguards.

*Further advice:*

## Watersports

All those undertaking watersports must comply with the DfEE recommendations on water confidence which say:

*"… everyone taking part can swim, in light clothing, a distance of at least 50 metres in the conditions likely to be encountered. The distance should be increased to 100 metres if the evidence is provided by a test in an outdoor swimming pool."*

All staff and young people must wear buoyancy aids to conform to BS.3595/69 or SBBNF/79. Staff should be proficient in life saving and artificial resuscitation techniques.

## Canoeing
*Staffing ratio* – Instructor controlled 1:6; Senior instructor controlled 1:8.
*Qualifications* – British Canoe Union: instructor, senior instructor (inland) senior instructor (sea), coach.
*Safety* – Canoes meeting BCU safety specifications, water conditions within capability of all concerned.
*Further advice:*

## Sailing
*Staffing ratio* – 1:6 in single handed boats.
*Qualifications* – RYA instructor's certificate/RYA senior instructor's certificate. (May supervise several craft at once).
*Safety* – Provide escort at all times to be able to move to scene of danger or accident on water immediately. Provisions to alleviate prolonged immersion in water.
*Further advice:*

## Sailboarding
*Staffing ratio* – 1:6.
*Qualifications* – RYA sailboard instructor.
*Safety* – Wetsuits, escort craft.
*Further advice:*

## Water skiing
*Staffing ratio* – 1:1 when learning.
*Qualifications* – British Water Ski Federation instructor/coach.
*Safety* – See all BWSF safety standards.
*Further advice:*

## Rowing
*Staffing ratio* – Circumstantial.
*Qualifications* – Amateur Rowing Association instructor.
*Safety* – Before self training must be proficient in capsize and accident procedures. See ARA.
*Further advice:*

## Underwater swimming
*Staffing ratio* – Circumstantial, pool, sea, age of group etc. 1:1 in training.
*Qualifications* – British Sub Aqua Club, snorkel instructor.
*Safety* – Extreme attention to all risk assessment factors: equipment, staff, group, water conditions, competence.
*Further advice:*

## Insurance

Before each trip the terms of the insurance cover and arrangements should be checked. There are six types of insurance that you should be familiar with. The officer responsible for insurance cover is ..................................... from whom further information can be requested.

### Third Party Public Liability Insurance

Policy No: ..............................................................................

Insurers name and address: ..........................................................

Officer responsible: ...................................................................

Indemnity level: £40,000,000.

This policy indemnifies youth and community workers or any party acting with the approval or consent, actual or implied of the county's youth service in respect of their legal liability to third parties (including young people and adults who are participating in youth and community activities) at any time and in any place (except when in a motor vehicle) when they are acting as specified within this document as a whole. Youth and community workers are effectively agents of the authority when young people and adults are in their care and should take every effort as specified in this document to take full and reasonable care.

### Employers liability insurance

Policy No: ..............................................................................

Insurers: ...............................................................................

Officer responsible: ...................................................................

Indemnity level: ......................................................................

This is used to indemnify the authority against claims from employees.

## Insurance for visits not involving overnight stays

Policy No: ................................................................................

Insurers:.................................................................................

Officer responsible: ...............................................................

The council provides special insurance for activities which do not involve an overnight stay. All trips in the category must be notified to the insurance officer responsible.

## Hazardous activities insurance

Policy no:................................................................................

Insurers:.................................................................................

Officer responsible: ...............................................................

Limit of indemnity:................................................................

This policy covers the following activities: Assault courses, recreation centres, activities in maintained sports centres, abseiling, go-karting, flying, gliding, motor cycling, Duke of Edinburgh, camping, caving, scuba diving, watersports, swimming, canoeing, cricket, football, fencing, hockey, mountain biking, war games, water skiing, wrestling, sub aqua, overlanding, mountaineering, rafting, rock climbing, ice skating, roller blading, parachuting, parascending, microlighting, hang-gliding, power boating etc..

All arrangements of the above activities must be notified in advance to the appropriate insurance officer.

### Visits and trips involving overnight stays

Policy No: ................................................................................

Insurers: ................................................................................

Officer responsible: ................................................................

Levels of indemnity: Medical and other £2,000,000; Luggage £800; Travel documents and replacements £200; Death £2,000 (all ages); Permanent total disability £30,000; Cancellation £20,000; Personal liability £1,000,000.

All trips involving an overnight stay must be notified in advance and checked with the appropriate insurance officer.

### Transport insurance

The county council provides a comprehensive range of insurance cover for all of its own vehicles used in youth and community service activities and for the drivers of such vehicles.

Policy No: ................................................................................

Insurers: ................................................................................

Officer responsible: ................................................................

Indemnity cover: ....................................................................

The county council makes no provision for the use of staff's private cars for the transport of users of the service and instructs all staff not to use their own car for the transportation of young people or adults participating in the activities of the service at any stage. The local authority has made provision for the insurance of staff's private cars in the case of accident, medical or other emergency.

## Further reading

Noble, Peter. *Risk Assessment and Safety on Education Excursions*, 1996, in conjunction with Rotheram Metropolitan Borough Council Department of Educational Services. Health and Safety Executive: *A Report into Safety at Outdoor Activity Centres*, 1996. Department for Education Safety in Outdoor Activity Centres: Guidance Circular 22/94 HMSO 1994. Adventure Education, *The Outdoor Source Book*, 1993. Smart, Julie and Wilton, Gill, *Practical Management: Educational Visits*, Campion Communications, 1995. Reed, Chris, *Red Light Spells Danger*, Youth Clubs, No.85, Autumn, 1996, pp 28-29. Putnam, Roger, *Safe and Responsible Youth Expeditions*. Young Explorers' Trust, 1994. Evans, Michael, *Safety on Educational Visits, A Quick Guide*, Daniels Publishing, 1995. Everard, Bertie, *Safety Principles, Codes of Practice and Value Statements*. Journal of Adventure Education and Outdoor Leadership, Vol 11, No 1, Spring 1994, pp 23-24. *Taking Groups Away*, St. John's Ambulance, 1992. *Safety Principles in Outdoor Education*, National Association for Outdoor Education, August 1987.

## Resources - training, organisations and publications

*Note:* This cannot possibly be an exhaustive list. I am mindful of the fact that youth and community projects cannot always afford the expensive publications and consultancies that relate to employment law and health and safety, so I have selected some particularly useful free or low cost resources. The key is access when required to paid time off for training. The TUC or GFTU basic courses are a good start, followed then by more advanced courses with the same organisations or the employer.

### The union

CYWU has a number of briefing papers on health and safety issues. The two most important topics written about recently are: *'Stress at Work'* and *'Violence at Work'*. The main manual that accompanies this pocket guide is also available from Russell House Publishing and outlines a three year strategy for developing improved health and safety practice in the workplace. The union intends to cover health and safety issues more frequently over the coming years through its national journal *Rapport* and has a specialist committee to progress this work.

CYWU membership means that workers should have a nearby health and safety representative. It also means that if staff are particularly interested in health and safety at work they have a chance to get elected to such a position themselves or to be selected to undertake the all expenses paid, very high quality health and safety training that the union can access for them.

CYWU also works in partnership with the National Union of Teachers whose health and safety department produces a range of informative leaflets which can be obtained through CYWU. An increasing number of youth and community workers are working in schools and colleges so the following leaflets through CYWU may be of interest. Among leaflets available through this route are: *Electricity in Schools, Eye Protection in Schools, Viral Hepatitis, Cavity Wall Insulation, Visual Display Units, Heating, Aids, Head Lice, First Aid, Dysentery, Administration of Medicines, Asthma; A Growing Problem, Pregnancy at Work, Crumbling Schools, Education (School Premises) Regulations 1996, Safety on School Journeys, Swimming, Harassment; A Union Issue, School Safety Representatives, Proposed Workplace (Fire Precautions) Regulations*. The Health and Safety Commission has also produced a useful publication *The Responsibilities of School Governors for Health and Safety* available from HMSO Bookshops. The Department for Education and Employment has also produced a range of publications on schools relating to health and safety issues, one of particular interest is *Improving Security in Schools*.

CYWU is affiliated to both the General Federation of Trade Unions and the Trades Union Congress and we receive all of their publications on health and safety. Local and national training can be accessed through these organisations.

CYWU's national office holds a small library of health and safety materials of particular interest in our sector and is open to members to visit by prior arrangement.

## The employer

If someone works for a local authority, their employer should have a health and safety committee which CYWU is part of and it should have an occupational health unit and a variety of training opportunities and publications.

The situation is less good in the voluntary sector, and staff may need to ensure that there is an extension of the partnership arrangements on other matters into the field of health and safety so that they can access local authority support. Many national voluntary organisations will issue their own guidance on health and safety and in some there will of course be expert knowledge of safety in outdoor education and so on.

The employer must provide workers with information, consultation, training and risk assessment.

## Key organisations

### Health and Safety Executive

This is an arm of government responsible for improving health and safety practice and having powers of inspection and enforcement. Look under Health and Safety Executive in the telephone directory to find the local HSE contact who may be approached for advice and a range of cheap or totally free publications. Among the most useful HSE publications for youth and community workers are: *Managing Health and Safety in Schools*, 1995; *Safety Policies in the Education Sector*, Revised 1994; *Workplace (Health, Safety and Welfare) Regulations*, 1992; *Guidance for the Education Sector*, 1995; *Violence to Staff in the Education Sector*, 1990.

In addition to the local contact you will find, the HSE resource and central office is as follows:

**HSE Infoline 0541 545500** – open 8.30am-5.00pm Monday to Friday.

There are three regional information centres:
- **Sheffield**, 0114 289 2333,
- **London**, Information Centre, Rose Court, 2 Southwark Bridge, London, SE1 9HS, and

- **Bootle**, St Hugh's House, Stanley Precinct, Bootle, Merseyside, L20 3QY.

Books and leaflets can be ordered from **HSE Books**, PO Box 1999, Sudbury, Suffolk, CO10 6FS, Tel: 01787 881165, Fax: 01787 313995.

The HSE has regional offices as follows:
- **Wales and South West:** Cardiff: 01222 263000, Bristol 01179 886000, Newcastle under Lyme: 01782 602300.
- **Home Counties:** Luton 01582 444200, Chelmsford, 01245 706200: London and South East: 0171 556 2100, Barking, 0181 235 8000, East Grinstead, 01342 334200.
- **Midlands Region:** Birmingham, 0121 607 6200, Northampton, 01604 738333, Nottingham, 01159 712800.
- **Yorkshire and North East:** Leeds, 0113 283 4200, Sheffield, 0114 291 2300, Newcastle Upon Tyne, 0191 202 6200.
- **North West:** Manchester, 0161 952 8200, Preston, 01772 836200, Bootle, 0151 479 2200.
- **Scotland:** Edinburgh, 0131 247 2000, Glasgow 0141 275 3000.

## Other organisations and publications

### ALARM
Association of Local Authority Risk Managers, Galaxy Building, Southwood Crescent, Farnborough, Hampshire, GU14 0NJ.

### Department for Education and Employment
The DfEE has worked on a lot of issues to do with school security and safety in the education sector. This work

carefully avoids any reference to the Youth and Community Service! However, before we persuade the department to do some specific work on our sector, it may be useful to consider some of the schools related work. The DfEE publication Guide 4, *Improving Security in Schools* is a useful starting point. There is also a School Security Team, DfEE, Sanctuary Buildings, Great Smith Street, London, SW1P 3BT, 0171 925 5000.

## Some building based support
The Department for Education and Employment also has an Architects and Building Branch which members lucky enough to be building new centres may wish to contact: Caxton House, 6-12 Tothill Street, Westminster, London, SW1H 9NF, 0171 273 6237.

*How to Combat Fire in Schools*, The Arson Prevention Bureau, 1993. APB, 140 Aldersgate Street, London, EC1A 4DD. This also contains some information that centre managers will want to consider. Although we note that fewer youth and community centres have been burned to the ground than other educational establishments.

*Building Security by Management and Design*, 1995, North East Risk Management Group, c/o West Denton High School, West Denton Way, Newcastle upon Tyne, NE5 2SZ.

*List of Approved Fire and Security Products and Services, A Specifiers Guide*, 1996. Loss Prevention Certification Board Ltd., Fire Protection Association, Melrose Avenue, Borehamwood, Herts, WD6 2BJ, 0181 236 9701.

## Labour Research Department
This is an independent research body that produces such high quality and inexpensive materials that CYWU

advocates that all members subscribe in their workplace and that all branches take out an affiliation. The LRD publications are indispensable for anyone interested in creating a more professional, healthier and safer working environment and for legal and good employment practice. Regular publications from LRD like their bulletin cover health and safety issues, but they also have separate pamphlets.

A brief but comprehensive and cheap key document is *Health and Safety Law a Guide for Union Reps.*

## NACOSS
National Approval Council for Security Systems, Queensgate House, 14 Cookham Road, Maidenhead, Berkshire, SL6 8AJ, 01628 37512.

## NCVYS
This is the umbrella body for voluntary sector youth organisations in England. Its sister organisations in Wales and Scotland are also detailed below. There are a range of specific and specialised publications.

## National youth agencies
There are three national youth work agencies, two are quangos and one is controlled by local authorities in England, though led by the voluntary sector. Health and safety issues are not features of their work but they all contain information centres which can provide information about work practice and work issues. This will frequently be vital in understanding the impact a work situation has on the health and safety of the professional worker.

## National governing bodies in some outdoor activities
Obviously there are a range of governing bodies for all sports, indoor and outdoor, which advise on standards and health

and safety matters specific to the activity. They set what is known as a matrix of competence for instructors and participants in their activities and this of course provides an essential backdrop to the risk assessment and evidence of skill in the pursuit. The telephone of some of the main organisations, particularly associated with outdoor adventure is given below. All sports and activities are regulated by a national body somewhere and your local library will be able to help you locate the main body concerned and its local and regional contacts.

British Canoe Union, 0115 982 1100,
Mountain Leaders Training Board, 01690 760350,
National Caving Association, 01749 870157,
British Cycling/Mountain Bike Federation, 0161 230 2301,
The British Horse Society, 01203 414288,
British Mountaineering Council, 01934 412295,
British Orienteering Federation, 01629 734042,
British Yachting Association, 01703 627400,
RYA Windsurfing, 01703 627400,
British Ski Foundation, 01506 884343,
The Ramblers' Association, 0171 582 6878,
Wales Mountain Leader Training Board, 01248 670964,
Scottish Mountain Leader Training Board, 0131 317 7200,
Scottish Canoeing Association, 0131 317 7314,
English Ski Council, 0121 501 2314,
Scottish National Ski Council, 0131 317 7280.

## ROSPA

The Royal Society for the Prevention of Accidents is based in Birmingham and has a comprehensive library. It is a membership based organisation. Membership gives access to its facilities and is therefore well worthwhile affiliating to at national voluntary organisation level or local authority level. ROSPA has many preventative and consciousness raising initiatives and is amenable to supporting various campaigns and pieces of guidance work.

### SSAIB
Security Systems and Alarm Inspection Board, 70/71 Camden Street, North Shields.

### The Suzy Lamplugh Trust
The National Charity for Personal Safety, 14 East Sheen Avenue, London, SW14 8AS, 0181 392 1839.

### TUC – National
A good source of advice and publications is of course the health and safety office. TUC, Congress House, Great Russell Street, London, WC1B 3LS, 0171 636 4030, fax: 0171 636 0632, e-mail: info&tuc.org.uk, telex 268 328 TUC G, website: http//www.tuc.org.uk

The Main TUC Publication, somewhat of a Bible is *Hazards at Work*.

### TUC – Regional TUCs
These can provide you with details of local health and safety training courses: Scottish 0141 221 8545, Northern 0191 232 3175, Yorkshire and Humberside 0113 242 9696, North West 0151 298 1225, Midlands 0121 622 2050, Southern and Eastern 0171 467 1291, Wales TUC 01222 372345, South West Region 0117 950 6425.

### TUC – Local
This is the local arm of the TUC, where trade unions come together. CYWU is affiliated to the TUC and therefore entitled to send delegates. Many Trade Union Councils have individuals or sometimes projects with great expertise in health and safety. Their addresses can be obtained from the National TUC or the Regional TUC.

## ACAS

The Arbitration Conciliation and Advisory Service is relevant in relation to health and safety as it can be used for advice on a variety of work related situations and it can also assist in disputes relating to the rights of representatives and workers. It is advisable to visit your nearest ACAS office and to be aware of their role and publications.

## Community Transport Association

The expert community minded national body that provides excellent publications, technical know how, conferences and general support.

## Environmental health officers

These are employed by local authorities and have rights of inspection and enforcement. They have considerable expertise and will prove a valuable source of support.

## Fire service

The local fire station and fire safety officer will of course be an invaluable source of information on fire regulations and fire prevention.

## Publications

### Journals and bulletins

*Rapport*
CYWU's bi-monthly national journal. Over the coming years will cover more health and safety issues appropriate to our sector.

*Labour Research*
Regular details of health and safety developments are contained in the monthly bulletin of this organisation, in addition there are a number of short, very user friendly

pamphlets on various aspects of health and safety.

*GFTU Research Service*
The General Federation of Trade Unions has considerable expertise in health and safety and produces bulletins kept by CYWU. Research into particular problems can be undertaken if requested by CYWU nationally.

*TUC Briefing Papers*
Many high quality and detailed reports on all aspects of health and safety. CYWU keeps these on file in its national office. To get them directly, contact the TUC's publications department for a complete reading list and order form: TUC, Congress House, Great Russell Street, London, WC1B 3LS.

*Adventure Education*
This frequently covers items related to outdoor education health and safety.

*Youth Clubs*
The National publication of *Youth Clubs UK*, frequently has articles of interest to health and safety especially in outdoor education work.

*Scouting*
As you would expect regular articles on matters relating to health and safety in hazardous activities.

*Adventure Activities Licensing Authority*
Tourism Quality Services Ltd, 17 Lambourne Crescent, Llanishen, Cardiff. Information on latest developments and standards can be supplied to those entered on the national mailing list.

## Other resources
This is a short list suggesting items that underpin much of the detail and which I believe will be most helpful in any

youth and community project, and cost effective. The items should be supplemented by as many of the free HSE publications as your project thinks necessary. They provide back up to assessing both the policies that the employer has in place and the quality of their attention to health and safety.

*Health and Safety, The New Legal Framework.* Smith, Ian, Goddard, Christopher, Randall, Nicholas. Butterworths, 1993. This is a comprehensive guide to the letter of the law, regulations and codes of practice etc.. The relationship between British legislation and European Directives is usefully explained also. It contains the text of all the relevant statutory instruments.

*Using Health and Safety Law. LRD's Guide for Safety Reps.* Labour Research Department. A forty page indispensable pamphlet.

*Hazardous Substances at Work: A Safety Rep's Guide.* Labour Research Department. Another short pamphlet that will greatly assist you in this area.

*Employment Practice and Policies in Youth and Community Work.* Nicholls, Doug. Russell House Publishing, 1995.

*Essentials of Health and Safety at Work.* Health and Safety Executive. £5.95

*Five Steps to Successful Health and Safety Management. Special Help for Directors and Managers.* Health and Safety Executive. Free.

*Five Steps to Risk Assessment. A Step by Step Guide to a Safer and Healthier Working Workplace.* Health and Safety Executive. Free.

*Negotiation Skills in the Workplace – A Practical Handbook.* Cairns, Larry. Pluto Press, 1996. The techniques of negotiation outlined in this publication will provide a good source of support when working to improve the workplace.

| **Useful contacts** | | |
|---|---|---|
| CYWU National Office, 302 The Argent Centre, 60 Frederick Street, Hockley, Birmingham, B1 3HS. | Telephone: 0121 244 3344 | Fax: 0121 244 3345 |
| Your local CYWU Health and Safety representative | Name: Tel: | Fax: |
| The officer responsible for health and safety in your employing organisation | Name: Tel: | Fax: |
| First aider covering your department/ building | Name: Tel: | Fax: |
| Doctor covering your building/project | Name: Tel: | Fax: |
| Your Occupational Health Department | Name: Tel: | Fax: |
| Health and Safety Executive (national) | Name: Tel: | Fax: |
| Health and Safety Executive (local) | Name: Tel: | Fax: |
| Local Environmental Health Service | Name: Tel: | Fax: |
| Local fire officer | Name: Tel: | Fax: |
| Regional TUC | Name: Tel: | Fax: |
| Regional TUC education officer | Name: Tel: | Fax: |

*Useful contacts – page 2*

| Local TUC | Name: | |
| | Tel: | Fax: |

| ACAS office | Name: | |
| | Tel: | Fax: |

| Outdoor education organisation | Name: | |
| | Tel: | Fax: |

| Health Education Authority | Name: | |
| | Tel: | Fax: |

Location of:

Accident report form ...........................................................

Health and safety policy ...........................................................

First aid box ...........................................................

Health and safety regulations ...........................................................

Health and safety inspection reports ...........................................................

Risk assessment documentation ...........................................................

Electrical testing reports ...........................................................